ADVANCES IN
MAN-MACHINE
SYSTEMS RESEARCH

Volume 3 • 1987

ADVANCES IN MAN-MACHINE SYSTEMS RESEARCH

A Research Annual

Editor: WILLIAM B. ROUSE
Center for Man-Machine Systems Research
Georgia Institute of Technology and
Search Technology, Inc., Atlanta

VOLUME 3 • 1987

JAI PRESS INC.

Greenwich, Connecticut *London, England*

55 Old Post Road, No. 2
Greenwich, Connecticut 06836

JAI PRESS LTD.
3 Henrietta Street
London WC2E 8LU
England

ISBN: 0-89232-659-X

Manufactured in the United States of America

CONTENTS

LIST OF CONTRIBUTORS

Sheldon Baron	BBN Laboratories, Inc. Cambridge, Massachusetts
John Seely Brown	Xerox Palo Alto Research Center Palo Alto, California
Richard R. Burton	Xerox Palo Alto Research Center Palo Alto, California
Keith D. Duncan	Department of Applied Psychology University of Wales Institute of Science and Technology
James D. Hollan	Institute for Cognitive Science University of California San Diego
Edwin L. Hutchins	Institute for Cognitive Science University of California San Diego
William B. Johnson	Search Technology, Inc. Atlanta
Timothy P. McCandless	Institute for Cognitive Science University of California San Diego
Nini Praetorius	University of Copenhagen
Mark Rosenstein	Institute for Cognitive Science University of California San Diego

Douglas M. Towne — Behavioral Technology Laboratories, University of Southern California

Louis Weitzman — Institute for Cognitive Science, University of California, San Diego

PREFACE

Man-machine systems is an interdisciplinary field of study that draws significantly on concepts and methods of psychology, systems engineering, computer science, management, and related disciplines. Because of this breadth, advances in the field emanate from a wide variety of sources and are published in a rather heterogeneous range of journals, monograph series, and so on. As a result, it can be quite difficult to assess the state of the art in man-machine systems. The goal of this annual series, *Advances in Man-Machine Systems Research,* is to contribute to ameliorating this difficulty.

Each volume in the series contains several substantive chapters that report new theoretical or empirical results, or novel applications, on topics such as human decision-making and problem-solving, human-computer interaction, human monitoring and control of dynamic processes, decision support systems, training systems, and basic human performance issues associated with these topics. Each chapter is a report of original research, typically involving the synthesis of a series of smaller efforts into an overall perspective on a topic. Chapters that primarily provide reviews of the literature are the exception rather than the rule, as are reports of any single theoretical or empirical study that could be published as a single journal article. Chapters that explicitly consider the applicability of the results reported to analysis, design, and/or evaluation of man-machine systems are particularly encouraged.

While there is no fixed plan to devote each volume to a special theme, the inaugural volume in 1984 dealt primarily with human decision-making and problem-solving in the operation and maintenance of complex dynamic systems. The second volume in 1985 focused on human performance in manual control tasks and, to an extent, decision aiding via adaptive models. This third volume emphasizes training systems in general and intelligent computer-based instruction and simulation in particular.

All of the chapters in this volume deal with operation and maintenance of technical systems. Electronic troubleshooting is discussed in the chapters by Towne, by Brown and Burton, and by Johnson. Marine power plants are considered by Hollan and his colleagues, as well as briefly by Johnson. Aircraft operations are discussed by Baron and aircraft maintenance by Johnson. Process control systems are considered by Duncan. Electric utility systems are discussed by Johnson. Thus, this volume clearly covers a wide range of technical training domains.

Of more importance, the chapters in this volume present reviews and summaries of a notable portion of the most outstanding research and applications in technical training over the past decade or two. While training technology has obviously been a key ingredient in these efforts, technology is not the primary issue in these chapters. Instead, educational philosophy, cognitive science, and implementation issues are the central themes.

Considering the six chapters in this volume as a whole, it is clear that microcomputers, videodisks, and large-scale graphic displays present tremendous opportunities for technical training. However, such opportunities do not guarantee success; rather, the crucial ingredient is understanding the learning process, from both individual and organizational perspectives, and embedding this understanding (implicitly or explicitly) in the learner-trainer system. This principle, of course, also applies to man-machine systems in general.

William B. Rouse
Editor

THE GENERALIZED MAINTENANCE TRAINER:

EVOLUTION AND REVOLUTION

Douglas M. Towne

ABSTRACT

The world is teeming with a vast array of specialized devices and equipments, all imposing their own unique behaviors on the people who must operate and maintain them. As with pilot training, simulators offer significant advantages over the actual systems for many phases of training. In maintenance training, however, it is not economically feasible to develop a unique simulator-trainer for each equipment to be taught.

In the late 1960s computer technology reached a level which made feasible the development of a "generalized" simulator–trainer that could be made to take on the characteristics of a particular equipment by changing software rather than hardware. For maximum ease of implementation, all the attributes of a particular

Advances in Man-Machine Systems Research, vol. 3, pages 1–63.

ISBN: 0-89232-659-X

equipment to be simulated could be expressed in a data base which could be produced by a nonprogramming subject matter expert.

This chapter describes the development of one of the earliest maintenance trainers to be developed to explore the efficacy of this philosophy, and it traces the evolution of this simulation system to a current maintenance trainer that provides highly individualized instruction as well as a refined and responsive simulation of system behaviors and internal system functions.

The chapter includes a full account of the expectations, successes, failures, detours, and surprises of this 20-year venture.

I. INTRODUCTION

Learning to diagnose and repair a complex system is a skill which is acquired and developed largely through practice. If the learner has received a firm grounding on the function, organization, and theory of operation of a system, practicing the diagnosis of real faults can both solidify information already studied and evoke new questions about the system. More importantly, practicing the diagnosis of unknown faults is the most effective way to learn to apply system knowledge to the performance of corrective maintenance.

A. The Problem

Surprisingly, some technical courses regard diagnostic skill as something that comes automatically with an understanding of system operation. Experience has shown rather clearly, however, that having knowledge of the system is necessary for acquiring diagnostic skill, but not at all sufficient. The learner must have an opportunity to practice applying his knowledge.

Unfortunately, it is difficult and expensive to provide meaningful and intensive diagnostic practice using the actual operational equipment. Generally, an instructor must be present to insert the practice malfunctions and to monitor the actions of each individual learner. While some group practice can be instructive, individual practice is also vital for students to become competent. As a result of these factors, the ratio of instructors to students is exceedingly high when the actual equipment is used for diagnostic practice.

Furthermore, the high cost and limited availability of operational equipment can essentially rule it out for training purposes. Typically, the operational needs for the system and its spare parts are so critical that training facilities are among the last to receive them. If the true cost of undersupporting training were fully recognized, this might not be the case. However, we would still seek more economical ways to provide realistic troubleshooting practice.

Cost, however, is not the only constraint on using operational equipment for training. The effectiveness of practice on the actual equipment can also be seriously limited. Typically there are only certain types of malfunctions which can be inserted into the real equipment without causing damage or undue delay. Furthermore, there may be limitations on the types of tests the learner is allowed to perform, on the modes of operation which can be configured, and on the quality of guidance obtained from busy instructors.

Many of these training difficulties can be overcome by the use of a computer-based simulator-trainer. If properly designed, a computer-based system can offer the student technician a wide range of self-paced self-administered diagnostic experiences which transfer well to the actual equipment.

But the lead time and cost to develop a dedicated and sophisticated training system can be prohibitive, too, especially if the target system is itself complex. Furthermore, much of this development effort is devoted to the implementation of training functions which are common across a wide range of applications; thus, replicating these functions in each dedicated trainer is inefficient.

For these reasons a "generalized" maintenance training system represents an attractive alternative. Such a system is domain-general; i.e., it contains no training detail which is specific to a particular target system. Instead, it accepts detailed and extensive data about a specific system and processes these specifications to deliver training which is specific to the target system and appropriate to moment-by-moment instructional demands.

B. Characteristics and Functions of a Generalized Maintenance Trainer

At Behavioral Technology Laboratories (BTL), we have developed a series of computer-based simulation trainers over a period of 20 years. The systems I will describe in this chapter are built around general-purpose computers that are capable of representing a particular system when supplied a data bank containing specifications for that system. The generalized maintenance trainer (GMT) of today performs three primary functions:

1. It simulates the operational equipment, in normal and failed modes, in responses to diagnostic actions by the student.
2. It acts as a surrogate instructor, selecting appropriate problems, providing within-problem guidance, and recording performance and progress data.
3. It constructs and maintains an internal model of the individual student, allowing it to tailor problem selection and guidance in an adaptive manner.

As we will see in the next section, our early trainers provided only the first of these three functions in their entirety and could better be termed *simulator* than *trainer*. But in a functional sense the more primitive systems were approximately equivalent to the Link trainer, which provided practice for new pilots; thus, the term *trainer* at least has some historical justification.

C. Advantages and Disadvantages of a GMT

It should be stated that the generalized trainer may not be the ideal solution to all equipment-training needs. There are situations which warrant the development of a unique training system to be used to train operation or maintenance of a specific target equipment or system. This could be true if there were a very large student population to be instructed in some single equipment or system. It might also be true if the ultimate operation of the actual system posed critical safety problems which could not be simulated with a generalized trainer. In this case the implications of improper operation may be so profound that no expense is spared in providing training.

It is more commonly the case that the number of people to be trained on a particular system is not enormous, the use of the system does not involve incalculable risk, and the cost of developing a special-purpose trainer is difficult to justify. In these cases, a generalized training system is feasible. Since 1968 I have been involved in developing an increasingly powerful succession of such systems. Each was designed to provide realistic diagnostic experience for a range of particular equipments and to avoid the expense of developing a special-purpose trainer.

As noted above, the GMT can provide greatly reduced cost *per application* in comparison to that of a number of dedicated systems. However, this benefit comes at a very substantial initial cost to develop the GMT capability. One might reasonably argue that if a GMT can do everything I have suggested, then there is no reason to develop more than one. To an extent, this is true. The latest version of GMT, described later, goes a long way toward providing simulation, instructor, and student functions that need not be developed again and that will offer intelligent instruction for a reasonably wide range of target systems.

Experience and realism, however, indicate that the development process is never completed. Continued research on human learning and understanding will certainly prompt numerous changes to the latest GMT, some of which will be easily incorporated into its existing structure, some possibly not. Further, computer technology will undoubtedly offer vastly improved processing capability and input–output media. Again, straightforward improvements in speed and memory would not render one version of a GMT obsolete, although they would open up new opportunities for machine assistance which could be added. Major improvements in display media and man–machine interaction techniques also could have major impact upon some portions of the GMT software architecture. Thus, development cost continues to be an issue in assessing the costs and benefits of a generalized approach.

Functionally, the distinguishing power of a GMT rests with the ease with which it can accurately take on the characteristics of particular systems. Some target systems, such as sophisticated radar units, may have displays which are not well described in a GMT specification. Other systems, such as nuclear power generation plants, may have functional behaviors whose complexity currently rules them out for application via a GMT.

The last issue to be discussed here is that of instructional power and consistency. Theoretically the dedicated trainer should hold some advantage over the generalized one, for very domain-specific instructional functions could be designed into the dedicated trainer. For example, a dedicated system might include a function which records, over a number of practice problems, the number of times the student performs tests in the high-voltage section of a system when the failure turns out to be in the low-voltage section. The trainer could then identify students which do not effectively discriminate between the two sections and could offer very specific remediation for that weakness. The GMT described in Section III offers mechanisms for specifying domain-specific "triggers" which sense and respond to particular qualities of a student's performance; however, there is little question that a dedicated system could be made more sensitive to particular aspects of the system than a GMT.

Practically speaking, the GMT may offer great instructional advantage over the dedicated trainer, for two reasons. First, enormous time and effort can be devoted to building effective pedagogical processes into a GMT. Additionally, with a wider community of users the GMT is more likely to receive the necessary field use to support subsequent refinement and enhancement. Secondly, the instructional processes applied by the GMT will be consistent across domains, whereas each dedicated trainer may operate under a somewhat unique instructional framework.

D. Purpose and Organization of the Chapter

This chapter will describe the development and evaluation of a family of generalized maintenance trainers, each of which was developed with the intent of improving upon and replacing its predecessor. The improvements from the early experimental systems have been profound and have resulted from research in instructional theory, from acquiring a deeper understanding of the ways people learn and retain information about devices, and from improvements in computer technology.

Section II of this chapter will trace the evolution of GMT systems with which I have been involved. The focus will be upon the things we learned from each version, the factors which allowed us to improve the previous system, and the ways we set out to develop a better system.

Section III describes current training technology in terms of the latest generalized training system. More specifically, the section deals with methods for simulating equipment function as well as appearance, for generating expert diagnostic performance, for selecting problems adaptively, and for interacting with the student.

Section IV considers some of the training research issues which have become pressing and pertinent, and Section V presents conclusions.

II. DEVELOPMENT HISTORY

This section traces the 20-year period of research and development leading to today's generalized, simulation-oriented maintenance trainer. The evolutionary process was characterized by a succession of systems, each representing the state of the art at the time of its original design. Typically the hardware configuration became partially obsolete shortly after completion, owing to the rapid changes occurring in the fields of computer technology. In some cases successes or failures in one system led directly to features in its successor. In other cases system capabilities were revised or added as a result of external developments in computer hardware, program development systems, or new understanding of human learning and knowledge representation.

A. Precursors to Generalized Maintenance Training

Two early developments provided key ingredients to the first generalized training system and have greatly influenced all the subsequent ones described in this chapter. They were—

- A quantitative model of the diagnostic process, termed BETS (Bayesian Electronics Troubleshooting System)
- A generalized representation of serial-action tasks, termed *Tasksym*

1. *The BETS Model*

The BETS model (Weis, 1963; Bond and Rigney, 1966; Rigney and Cremer, 1966) is a formulation of the diagnostic problem in information theory terms, such that one can (1) compute the expected value or benefit of a test under consideration and (2) interpret the significance of specific test results.

The model applies Bayes' theorem (Edwards, 1962) of statistical inference to the troubleshooting domain. This theorem expresses the probability that a particular hypothesis is true, following the receipt of some new data, in terms of the conditional probability of obtaining that data given that the hypothesis is true and the probability of the hypothesis prior to receiving the data.

For diagnosis, the hypothesis is the proposition that a particular fault exists in a system, and the data are results from tests. Application of Bayes' theorem to fault diagnosis requires knowledge of the effects of the possible failures, as might be contained in the symptom-malfunction (SM) matrix shown in Table 1.

Suppose we have no knowledge of the likelihoods of the four failures in the system, in which case the four (a priori) failure probabilities are all 0.25. If we perform a test by checking indicator 2, and see that it is ON, then according to Bayes' theorem the new (a posteriori) probability that the failure is failure 1 is

$$p_1 = P(H_1|\text{ON}) = \frac{p(H_1)P(\text{ON}|H_1)}{\Sigma P(H_j)P(\text{ON}|H_j)}$$
$$= \frac{0.25 \times 1}{(0.25 \times 1) + (0.25 \times 0) + (0.25 \times 1) + (0.25 \times 1)} = 0.33.$$

Similarly, the probabilities of failures 3 and 4 are each 0.33, while the probability that failure 2 is true is 0.00.

Table 1. Simple Symptom-Malfunction Matrix

	INDICATOR		
FAILURE	1	2	3
1	100	ON	RED
2	0	OFF	RED
3	100	ON	BLUE
4	50	ON	BLUE

For this simple example, all initial probabilities are equal and each failure mode produces a unique, nonprobabilistic symptom at each indicator. As a result, each new test result completely eliminates one or more hypotheses until only one is left. In a more realistic case, the initial suspicion levels are not equal, as some reliability knowledge is often available, and the possible effects of failures may be probabilistic in nature. Bayes' theorem deals properly with the realistic situation.

As a model of human diagnostic behavior, Bayes' theorem has several serious shortcomings. First, studies have shown (Edwards, 1968; Manz, 1970) that people's subjective probabilities, or beliefs, depart in significant and systematic ways from the predictions of the theorem; i.e., their beliefs following the receipt of new information do not correspond closely with the Bayesian formulation. These biases alone would not spell disaster for Bayes' theorem as a model of human behavior, for they could be incorporated into a revised model.

Unfortunately, our early studies (Rigney et al., 1967) showed that technicians did not have nearly enough accurate fault-effect knowledge to operate in a Bayesian manner, or some revised Bayesian manner, for the circuits tested. The studies showed, however, that those few individuals who thoroughly understood the possible failure modes of a system behaved in ways which were generally consistent with Bayes' theorem, in terms of their relative suspicions of various failures.

Since those early studies employed simple electronic circuits as experimental vehicles, however, our conclusions were that technicians were not likely to perform like Bayesians when troubleshooting complex systems consisting of hundreds or thousands of such circuits. As a result, we abandoned the notion that Bayes' theorem could be a useful model of human performance, but we retained the computation as a rational means for interpreting the significance of test results. This was an important part of assessing the expected value of alternative tests under consideration, as described next.

The second part of the BETS model was the application of the classical information theory measure of uncertainty, or entropy, to computing optimal testing sequences (Johnson, 1960). According to this formulation, the uncertainty about the true state of a system can be computed as follows (all logarithms are to the base 2):

$$U = \Sigma p_j \log p_j$$

where p_j is the probability that failure j exists and U is the uncertainty about the true failure condition of the system.

For a system consisting of 10 equally reliable components, each of which can fail in just one mode, the system uncertainty is

$$
\begin{aligned}
U &= 0.1 \log 0.1 + 0.1 \log 0.1 + \cdots + 0.1 \log 0.1 \\
&= 10\,(0.1) \log 0.1 \\
&= \log 0.1 \\
&= -3.322
\end{aligned}
$$

There is higher (absolute) uncertainty concerning the state of a system of 10 equally likely failures than one with 5 equally likely failures. Further, the uncertainty for a particular system is greatest when all the probabilities are equal. The minimum uncertainty is achieved when one of the failure probabilities reaches unity and all others are zero.

Now, using the symptom-malfunction matrix described above, one can compute, for each test under consideration, the possible outcomes of that test and the probabilities of each of those outcomes (since the probabilities of the individual failure hypotheses are known). Then, using Bayes' theorem one can compute the probabilities, or suspicion levels, which would result following the test. From this can be computed the *expected* system uncertainty which will result *if the test is performed.*

By doing this for all possible tests, one can determine the relative expected value of each test. Based upon this, one can evaluate the wisdom of performing some proposed test, and one can recommend a better test if necessary. These two functions would be central to the generalized maintenance trainers to follow.

2. *Tasksym*

While the BETS model provided a powerful basis upon which to evaluate a student's diagnostic action selections, there remained the problem of evaluating the correctness of procedural work. How, for example, could a complex task be represented in such a manner that a student's sequence of actions could be compared to it? To do this required either (1) enumerating all feasible action sequences which attained the task goal or (2) representing tasks in a way which implied all the correct alternative sequences. The former technique is unworkable for all but the simplest of real-world tasks. The latter approach was taken in a related research effort (Towne, 1969), which had as its goal the development of a generalized technique for computing optimal action sequences for attaining procedural goals.

Based on earlier work by Slagle (1963), I developed a number of simple operators which corresponded to constraints on the performance of actions. Three major operators were (1) SEQ, which specified that the actions in its argument list must be performed in sequence; (2) ANY, which specified that the current goal would be attained if any one of the alternative actions were performed; and (3) ALL, which required the performance of all actions in the list, in any order.

For example, a simple task requiring the performance of actions A1, A2, and A3 in order would be represented as the following list:

(SEQ A1 A2 A3)

This simple task could be a subgoal within a larger task. If it was one of several subgoals which must be attained (in an arbitrary sequence) to satisfy the larger goal, the larger task might be specified as follows:

(ALL (ANY A8 A12 A54) (SEQ A1 A2 A3)

In this fashion tasks of arbitrary complexity could be represented. A graphic representation of a simple task, that of changing an automobile tire, is shown in Figure 1. Figure 2 presents the lists which specify the task in Tasksym format.

Among the functions developed in Lisp were the following:

- A function for computing the optimal way to perform the task, to minimize performance time (this function recognized sequence-dependent constraints, which affected total task performance time)
- A function for evaluating the correctness of any completed or ongoing performance sequence
- A function for diagnosing a student's knowledge of a task
- A function for explaining the performance of a task, in terms of its goal structure

These functions all operated upon lists specifying particular tasks and set the stage for generalized representations of equipment behaviors. In addition to the lists of goals to be attained, I developed lists which specified negative events, which I termed *antigoals*. An antigoal is a condition to be avoided, such as coming in contact with a high-voltage source and ground or energizing an equipment with the safety interlock disabled. The structure of an antigoal is represented in exactly the same format as the primary task goals.

For one system to which I applied this notion, the AN/URC-32 radio transceiver, I found that the correct procedure for tuning the transmitter brought the technician perilously close to also attaining a major antigoal, that of blowing out the transmitter. This motivated the development of one final Lisp function in Tasksym whose purpose was to monitor the attainment of subgoals within the antigoal structure and alert the user (soon to be a student) when attainment of the antigoal loomed close. The concept of antigoal was useful for at least two reasons: (1) it allowed for the explicit specification of negative events, thereby supporting instruction in the conditions to be avoided; and (2) it provided a way to specify exceptions to general goals, thereby avoiding the necessity of con-

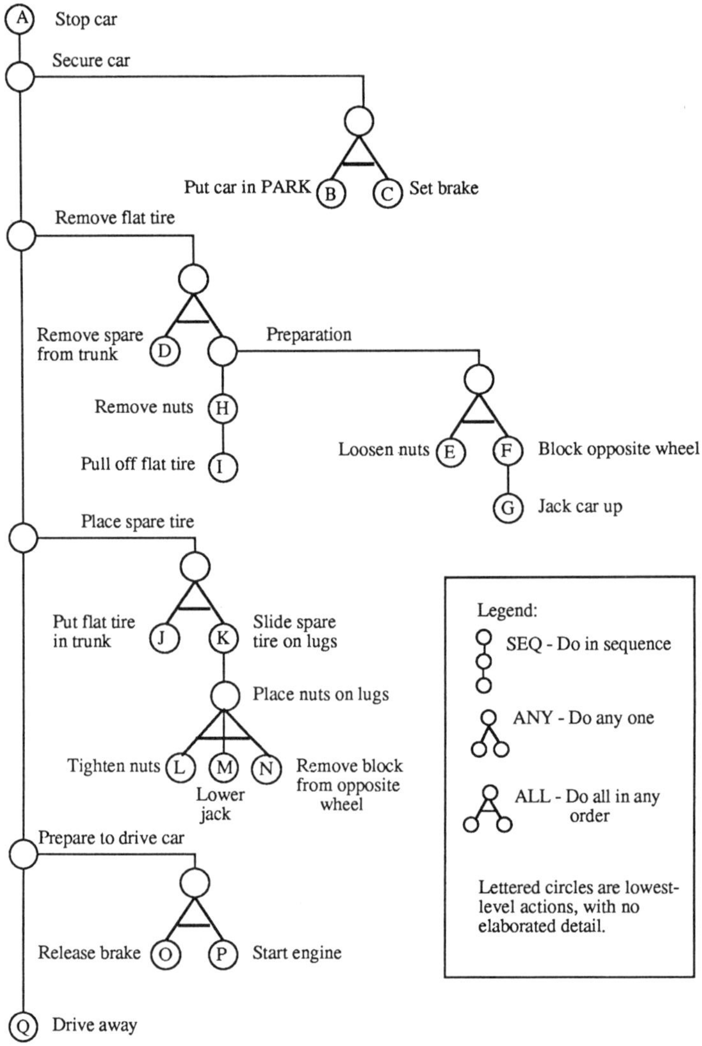

Figure 1. Graphic representation of task structure for CHANGE TIRE.

structing extremely complicated goal specifications which implicitly exclude the negative event.

B. TASKTEACH: An Early Generalized Maintenance Trainer

Between 1968 and 1972 I worked with Joseph W. Rigney, Carole A. King, Robert Cremer, and Nicholas A. Bond, to develop TASKTEACH, possibly the

```
(SET 'CHANGE TIRE' (SEQ ('STOP CAR'  'SECURE CAR'  'REMOVE FLAT TIRE'
                         'PLACE SPARE TIRE'  'PREPARE TO DRIVE CAR'  'DRIVE AWAY')))

(SET 'SECURE CAR' (ALL ('PUT CAR IN PARK'  'SET BRAKE' )))

(SET 'REMOVE FLAT TIRE' (ALL ('REMOVE SPARE FROM TRUNK'
                         (SEQ (PREPARATION  'REMOVE NUTS'   'PULL OFF FLAT TIRE')))))

(SET PRAPARATION (ALL ('LOOSEN NUTS' (SEQ ('BLOCK OPPOSITE WHEEL'  'JACK CAR UP')))))

(SET 'PLACE SPARE TIRE'  (ALL 'PUT FLAT TIRE IN TRUNK'
                         (SEQ ('SLIDE SPARE TIRE ON LUGS'  'PLACE NUTS ON LUGS'
                         ALL ('TIGHTEN NUTS'  'LOWER JACK'
                              'REMOVE BLOCK FROM  OPPOSITE WHEEL')))))))

(SET 'PREPARE TO DRIVE CAR'  (ALL ('RELEASE BRAKE' 'START ENGINE')))
```

Figure 2. Input lists specifying the serial task of changing a tire.

Table 2. Summary of Generalized Maintenance Training Systems

	Maintenance Training System					
	TASK-TEACH I	TASK-TEACH III	GAIN	GMTS	ESAS	IMTS
Development period	1968-1970	1970-1972	1972-1974	1978-1983	1980-1983	1984-1986
Computer	Time-shared	Time-shared	IMLAC graphics terminal	TERAC 8510	IBM XT/AT	XEROX 1108 LISP Machine
Program language	LISP 1.5	Fortran IV	Assembly language	UCSD Pascal	TenCORE	Interlisp-D
Graphic medium	None	Color slides	Color slides	Color microfiche	Computer graphics	Computer graphics
Touch input	None	None	Light pen	Sonic pen	Mouse	Mouse
System model	None	None	Connectivity model	Indicator state rules	Indicator state rules	Functional model
Diagnostic guidance	Based on optimum	Based on optimum	Based on optimum	None	None	Based on expert
Problem selection	Student controlled	Student controlled	Student controlled	Random from sets, or by instructor	Student controlled	Adaptive

first generalized computer-based maintenance training system. Table 2 lists the characteristics of TASKTEACH, along with those of the other systems which will be described.

The design of TASKTEACH was influenced by many researchers; the concepts of Glenn Bryan of the Office of Naval Research were especially compelling and found their way into the resulting system in many forms. Joseph Rigney, with whom I worked over the next 12 years, was one of the first to recognize the potential of computer-based systems for instructional purposes and was a leader in identifying the types of research which would be necessary to support true progress in the field.

The philosophy we shared which motivated our approach was (1) that the student might have something to contribute to the teaching–learning process and (2) that machines might be fully capable of offering a permissive and supportive environment for learning. It was argued that, contrary to the programmed instruction technique which was then the rage, computers might allow the learner to influence the instructional process, as they do in reasonably sized conventional classrooms, and that students might be more effective if they know they can make a mistake without being routed to a punishing remediation phase.

As a result of this thinking and the availability of the BETS model and

Tasksym system, TASKTEACH was developed to offer the following types of maintenance training for any specified system:

- "Front-panel" drill, involving the location and identification of switches, sections, indicators, test points, etc.
- Guided practice performing serial action maintenance tasks, with provisions for avoiding antigoals; the instruction included diagnosis of the student's entering capabilities via an interactive checkoff of the subtasks already mastered.
- Guided practice performing fault isolation of computer-selected faults; this included part-task exercises in which the learner judged the normalcy of various test results and predicted symptoms from known failures (which we termed *backward troubleshooting*). Recommended tests were selected according to the optimal information-theoretic model described above, and student-performed tests were evaluated against this reference.

The original version of TASKTEACH (Rigney, Towne, and Bond, 1969) was highly student-controlled, in contrast to the relatively rigid training experience which could be provided in the military training setting, using real equipment. Using typed command words, the student could review completed work, request explanations, check progress in relation to task goals, and request help in selecting the next action.

The notion of assigning some responsibility and associated control to the learner was met with dismay by some and high regard by others. At the time we talked of adaptively assigning this responsibility and control to the learner and to the executive instructional programs depending upon individual performance, but we really had no generalizable theory upon which to base such an allocation. It was clear, however, that providing considerable control to the student created quite a different learning environment than one in which all training decisions are made by the training system. To some students the increased control was a unique opportunity to explore; to others it was an increased burden for managing their own learning. In any case, high student control was a virtual necessity at the time, as it allowed the exploration of generalized training techniques by avoiding the necessity of modeling the student, which was then more than we were prepared to undertake.

Figures 3–5 provide sample listings of the earliest student–computer interactions with what is now called TASKTEACH I. Figure 3 shows a procedural drill in which the student performed work on actual equipment or a mock-up while viewing a slide showing the structure of the task, as in Figure 1. Figure 4 shows a front-panel drill session in which the student locates controls and indicators named by TASKTEACH. Figure 5 is a sample troubleshooting exercise in which TASKTEACH checks the student's readings and conclusions as the student performs tests on a failed circuit.

THIS IS PROGRAM TASKTEACH	
THE FOLLOWING COURSES ARE AVAILABLE	
1. USE OSCILLOSCOPE 2. FRONT PANEL DRILL-OSCILLOSCOPE 3. TROUBLESHOOT PHASE-SHIFT OSCILLATOR	
ENTER THE NUMBER OF THE TASK YOU WISH TO DO: **1** MODE? **1**	(This provides maximum support.)
THE TASK IS DISPLAYED STARTING ON SLIDE 1	(This shows a diagram of the task, as in Figure 1)
REPORT EACH ACTION YOU TAKE. * **RCCW** * **PON**	(These are codes for controls and settings, from the diagram.)
* **NEXT**	(This is a request for help selecting the next action.)
DO ANY OF THE FOLLOWING: SON CSA	

* **CSA**
* **SON**

THE COMPLETED GOALS NOW ARE:
SWEEP SET-UP

(TASKTEACH advises the student whenever goals are attained.)

* **NEXT**

DO ANY OF THE FOLLOWING:
VSA20

* **VSA20**
* **BFDEP**
* **R**
* **HP**
* **VP**

THE COMPLETED GOALS NOW ARE:
SWEEP SET-UP
CENTER THE BEAM

* **R**

SORRY, R IS NOT CORRECT

* **BFREL**

(The student has either mis-read the task structure slide, or has used the wrong control.)

etc.

Figure 3. TASKTEACH procedural drill (student entries in bold).

THIS IS PROGRAM TASKTEACH

THE FOLLOWING COURSES ARE AVAILABLE

1. USE OSCILLOSCOPE
2. FRONT PANEL DRILL-OSCILLOSCOPE
3. TROUBLESHOOT PHASE-SHIFT OSCILLATOR

ENTER THE NUMBER OF THE TASK YOU WISH TO DO: **2**

I'LL GIVE YOU THE CONTROL NAMES. YOU ENTER THE CODES.
TYPE A QUESTION MARK IF YOU CAN'T FIND THE CONTROL.

HORIZONTAL DISPLAY IS: **HD**

(Here the student locates each control named by TASKTEACH and reads off the code placed there by the instructor. This could involve either the actual equipment or slides.)

TRIGGER SOURCE IS: **TO**
WHAT? **TSO**

CALIBRATOR IS: **C**

POLARITY, CHAN B IS: **PA**
NO, THAT'S THE POLARITY, CHAN A
TRY AGAIN: **PB**

(The student locates a wrong control.)

VERTICAL SENSITIVITY VERN; CHAN A IS: **?**
THE VERTICAL SENSITIVITY VERN; CHAN A IS CODED **AV**

(The student can't find the control.)

etc.

Figure 4 TASKTEACH front-panel drill (student entries in bold).

These sample listings reflect the tremendous constraints which the primitive computer hardware of the era placed upon the instructional interactions. The original TASKTEACH program was run on a time-shared computer linked by telephone to a teletypewriter, printing at the rather languid rate of 10 characters/sec. The student was required to learn command words to communicate with the system and had to refer to external figures (slides) or mock-ups for codes placed there for reference. The requirement that all interaction be in textual form added typing and memory activity to the task of learning.

Because of the serious limitations in interactive power, TASKTEACH could offer only two modes of practice, neither of which were entirely satisfactory. In one mode (as in Figure 5), the actual equipment was used for the setup drills and for practicing taking measurements with test equipment. Since there was no link between the real equipment and the computer, the student reported results to TASKTEACH via the keyboard and received typed responses. This mode had the great disadvantage that someone had to insert the malfunction into the equipment prior to each exercise. In the other mode, the student worked from diagrams and pictures of the actual equipment, which often forced the instruction to be *about* the tasks rather than involving performance of the tasks themselves. In one study a full-scale mock-up of an equipment front panel was used, with relatively satisfactory results.

After working long hours at the terminal, students completed a questionnaire to register their impressions. Surprisingly, students judged the learning experience as interesting, in spite of the slow and laborious techniques required for interaction, and they gave good marks to the quality of learning which they experienced. The most common responses were that TASKTEACH exposed them to more possible malfunctions and it related front-panel symptoms to internal circuitry involved. The most common criticisms were that (1) they could not see the equipment being discussed and (2) they would prefer to request the review of some completed problems rather than being forced to review each one. Thus the students were expressing their preference for even greater control over the instruction.

The test records confirmed that they had significantly improved their abilities to perform the maintenance tasks exercised. Average completion time per problem dropped from 16.6 min during practice sessions to 13.5 min on test problems; solution rates increased from 83 percent to 98 percent; and front-panel errors dropped from 34 percent to 15 percent. For reasons never fully understood, the average number of replacements per problem increased from 1.27 during practice problems to 1.58 on test problems. Perhaps the practice problems were rendered easier than the test problems by the availability of the support functions.

Of course these results said nothing about the relative rate of learning which occurred on the trainer compared to equal time in a classroom or working on the actual equipment. We had learned, however, that generalized systems could

THIS IS PROGRAM TASKTEACH	
THE FOLLOWING COURSES ARE AVAILABLE	
1. USE OSCILLOSCOPE	
2. FRONT PANEL DRILL-OSCILLOSCOPE	
3. TROUBLESHOOT PHASE-SHIFT OSCILLATOR	
ENTER THE NUMBER OF THE TASK YOU WISH TO DO: **3**	
DO YOU WANT ME TO SELECT THE MALFUNCTION? **YES**	
SHALL I PROVIDE THE READINGS? **NO**	(The student wants to practice troubleshooting, taking actual readings on the equipment.)
TP: **7**	
TYPE: **DC**	
READING: **145**	
RIGHT. IS THIS H, N, OR L? **H**	(The student judges the reading to be Higher than normal.)
RIGHT.	
TP: **3**	
TYPE: **R**	
READING: **160**	
RIGHT. IS THIS H, N, OR L? L	(The student believes incorrectly that 160 is Lower than normal.)
NO, THINK AGAIN. IS THIS H, N, OR L? **N**	
RIGHT.	
.	
.	
.	
TP: **MAL**	(The student believes he has identified the malfunction.)

ENTER THE NUMBER OF THE MALFUNCTION YOU SUSPECT: **2**	
SORRY, R1-SHORT IS NOT THE PROBLEM.	
TP: **PROG**	(The student requests a PROGRESS report.)
REMAINING POSSIBLE MALFUNCTIONS ARE	(These are the malfunctions which the student's tests have not yet eliminated.)
R2-OPEN	
R2-SHORT	
R4-OPEN	
R5-SHORT	
TP: **NEXT**	(The student asks which test would be the best one to perform next.)
THE BEST NEXT TEST IS TEST POINT 7 , AC	
TP: **2**	
TYPE: **R**	
READING: **0**	(Here, the student has made an error using the test equipment.)
SORRY, 0 IS WRONG. TRY AGAIN.	
READING: **500**	
RIGHT. IS THIS H, N, OR L? **N**	
RIGHT.	
TP: **MAL**	
ENTER THE NUMBER OF THE MALFUNCTION WHICH YOU SUSPECT: **3**	
THAT'S RIGHT, THE MALFUNCTION WAS R2-OPEN	
HERE'S A NEW PROBLEM.	
etc.	

Figure 5. TASKTEACH troubleshooting exercise (student entries in bold).

provide an effective environment for learning and that TASKTEACH could provide a learning experience which students might not otherwise receive in conventional training.

By 1972 TASKTEACH had been converted from Lisp to Fortran IV, so that it could be run on a local time-shared computer (Rigney, Towne, King and Langston, 1972). By this time the student station consisted of a 300-baud alphanumeric cathode-ray tube (CRT), with a random-access (but student-controlled) slide projector. This version of TASKTEACH (by now version III) provided varying levels of support for all three types of instruction (procedural drill, front-panel drill, and troubleshooting exercise). Level 1 support was a closely supported mode in which the learner was led through new procedures and practice problems. Level 2 was the practice mode, where the learner initiated actions on his own but received notice of serious errors and remediation as required. Level 3 was a self-test mode, while level 4 was a test for the record. The selection of level was under student control, although poor performance at a level could evoke a recommendation to return to a higher support mode.

Unlike TASKTEACH I, this version of TASKTEACH could deal with student actions which changed the setup of the equipment (recall that TASKTEACH I only dealt with troubleshooting of electronic circuits or other units of equipment which did not change configuration). Although the mode of interaction was still alphanumeric, the student could now operate eqipment, report each action to the instructional system, and obtain assistance in performing fault diagnosis.

Figure 6 presents a sample listing of a diagnostic drill. In this example entries such as I2 indicate that the student wishes to view indicator 2; TASKTEACH then looked up the symptom which the assumed malfunction produced at the indicator and would report that result back to the student. An entry such as S2-1 indicated that the student would like to set switch 2 to position 1. TASKTEACH assumed this action had been performed and updated its internal image of the state of the equipment. This internal representation was nothing more than a list of the current switch settings. Since TASKTEACH had no model of the actual equipment, it worked from complete enumerations of the possible switch setting combinations. While this worked satisfactorily for simple systems, it would clearly not suffice for systems having many switches.

At this time the first blush was wearing off "computerized" instruction. The cost per hour of connect time was still excessively high, and many considered that only highly paid pilots and executives would make suitable students for this expensive approach. We reasoned, however, that time sharing would proliferate as the use of computers spread, just as our national highway network expanded to serve the automobile. We saw the day when satellites would bring cheap connect time to every household and therefore forged ahead with software development on the premise that time-sharing costs would drop substantially over the next three to five years. We resolved to be ready with the software when the capabilities and economy of the hardware caught up.

ACTION(S)? **PROG**
THERE ARE (52) POSSIBLE MALFUNCTIONS REMAINING

ACTION(S)? **I1,I2**
I1: VIDEO SWEEP IS NORMAL
I2: CURSOR SWEEP IS NORMAL

ACTION(S)? **PROG**
THERE ARE (49) POSSIBLE MALFUNCTIONS REMAINING

ACTION(S)? **S2-2,I16**
S2-2: NORM-ALIGN SWITCH S2 NOW SET TO ALIGN
I16: ALIGN VIDEO BEAM IS NORMAL

ACTION(S)? **S2-1,I16**
S2-1: NORM-ALIGN SWITCH S2 NOW SET TO NORM
I16: ALIGN VIDEO BEAM SHOULD NOT BE CHECKED
IN THIS CONFIGURATION.

ACTION(S)? **S2-2**
S2-2: NORM-ALIGN SWITCH S2 ALREADY SET TO NORM.

ACTION(S)? **REV**

ENTER THE MALFUNCTION(S) WHICH YOU SUSPECT? **M51**

M51: A5C9 SHORTED; OR K8B OPEN CONTACT 5 COULDN'T
BE THE MALFUNCTION.
CHECK YOUR STEP NUMBER 1

FOR SOME REASON, YOU ARE NOT SUSPECTING THE
ACTUAL MALFUNCTION.

THERE ARE 48 OTHER MALFUNCTIONS WHICH COULD
CAUSE THE INDICATIONS YOU HAVE RECEIVED.

etc.

Figure 6. TASKTEACH diagnostic drill (student entries in bold).

Stated another way, we made exactly the right decisions, for exactly the wrong reasons. Per-hour cost of computers fell drastically, of course, for a reason we did not recognize when we first saw it. The reason for this revolution in computer use first came to our notice in the form of a new ''programmable graphics terminal.'' Manufactured by Imlac Corporation, the PDS-1 terminal could display graphics as well as text. To minimize transmission time, the terminal could be microcoded to produce its graphics locally, from higher-level commands coming from the central computer. A tiny central processor in the terminal was actually capable of performing limited computer-like functions to make this all possible.

C. GAIN: The First Stand-alone System

By the time we had secured possession of a PDS-1 the manufacturer had doubled the size of the memory, allowing small programs to be downloaded from the main computer and to be executed locally. Now we revised our prophecies: programmable terminals would allow expensive time-sharing costs to be reduced by going off-line when the main computer was not needed. In late 1973 Joe Rigney and I said the following (Rigney and Towne, 1974.)

> Some of the possible configurations of hardware and software for CAI [computer-aided instruction] systems range from large networks with one or more large computers to small, stand-alone systems based on minicomputers. The latter have some advantage for use in certain types of environments. For example, where the subject-matter is classified and there are not sufficient local resources to sustain a larger time-sharing configuration with central [central processing unit] CPU and peripheral terminals.
>
> We doubt that ''either–or'' arguments for these two extremes are appropriate. The variables to consider are complex and there probably is room in the world for several different configurations. A small programmable graphics terminal recently became available on the market for use as a stand-alone CAI system. This device is relatively inexpensive, yet is surprisingly powerful if its inherent capabilities are recognized and are fully exploited. It also could be a forerunner of even more powerful and less expensive programmable graphics terminals made possible by the remarkably swift advances in integrated circuit, microprogramming, and memory technologies.

Thus we began work on a new generalized maintenance trainer, one which would run on the PDS-1 programmable terminal and would allow graphic displays and light pen inputs.

Like all previous systems we christened this descendent of TASKTEACH; it was called GAIN, for Graphical Analysis of Interactive Networks (Rigney, Towne, and King, 1974). Development of GAIN was painfully slow, owing to the necessity of programming in assembly language, a torture I had avoided since completing Computer Programming 101. But the fascination of producing a pedagogically powerful interaction between student and instructional program was sufficient to sustain the effort. Unlike its predecessors, GAIN dealt with a visual representation of the actual system, either in terms of front-panel element appearances or internal functions in the form of block diagrams. Now both the student and the instructional program could point to and alter front-panel switches, they could trace signals on the screen, they could point out suspected blocks, and they could signify what blocks would be monitored by any test.

By the time GAIN was completed, the PDS-1 programmable terminal had become a complete stand-alone minicomputer. While it was still capable of communicating with a larger host and was still regarded as a local processor to be used in conjunction with a more powerful computer, we ceased to think of it in any terms except as a fully stand-alone and independent training system.

1. *Physical Simulation*

Because the highest-level graphic command in this early graphics computer was a vector line-drawing command in assembly language, creating graphic images was both slow and dependent upon programming skills. As a result, I developed a primitive graphic editor which allowed nonprogrammers to create new images of switches and indicators. Even this was a tedious task, however. Figure 7 is an example of the initial graphic presentation displayed by GAIN when simulating equipment front panels. In this example the target equipment was the AN/URC-32 transmitter–receiver. In GAIN the graphic screen was split into two functionally identical halves. The student could select any subunit in

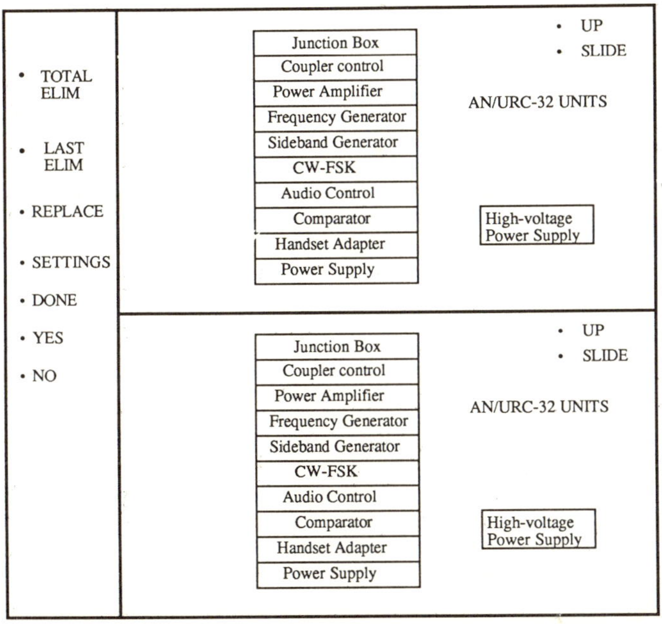

Figure 7. Initial GAIN display of AN/URC-32.

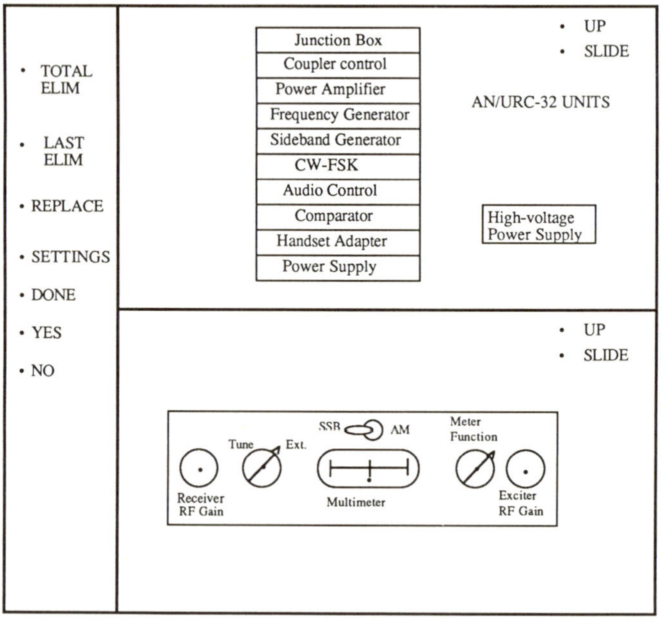

Figure 8. Display of sideband generator unit in lower window.

either half with the light pen, causing the original display to be replaced by a more detailed representation of the selected unit. In this way, the student could obtain closeup views of two front-panel elements such as a switch and an indicator. Only those closeup views of individual front-panel elements provided displays which revealed the states of the elements. Upon touching the light pen to the sideband generator unit, in the lower screen half, the student would see a display as shown in Figure 8. By then selecting the meter function switch from the lower display and similarly stepping down to the multimeter in the upper display, the student would see the display shown in Figure 9.

The menu down the left-hand side of the screen displayed the commands which allowed the student to "replace" a suspected component (REPLACE), to see a text listing of all current switch settings (SETTINGS), to check on the correctness of his test interpretations (TOTAL ELIM and LAST ELIM), and to claim that a problem was resolved (DONE).

2. *Functional Simulation*

Initially, block diagrams were entered into GAIN for purposes of communicating about the internal sections of the system being simulated. In this manner the student could point out sections believed to be confirmed as operational, he could offer his understanding of the blocks monitored by various tests, and GAIN

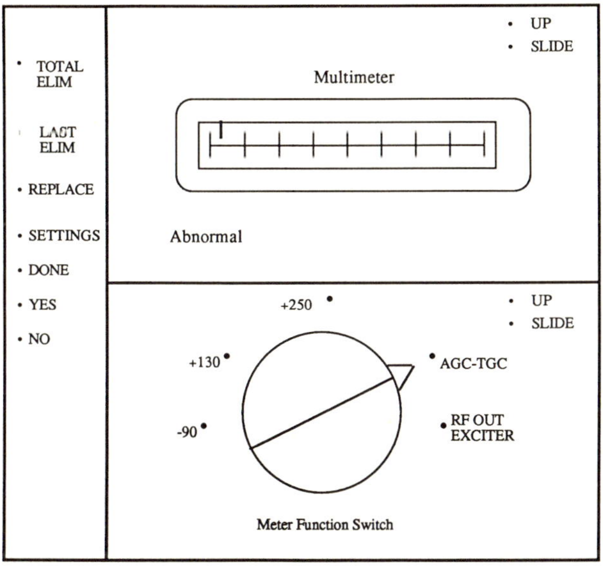

Figure 9. Lowest level display in both windows.

could highlight certain sections in response. Soon it became apparent that the graphic display offered far greater potential than was being utilized. We had to stop regarding the computer screen as a substitute for a static schematic diagram on a piece of paper. We came to recognize that the graphics capabilities should be utilized to respond to the learner's actions in ways which were functionally like the ways the real equipment would respond. The student should be able to operate the graphic representation the way he does the actual system, and the graphic representation should respond accordingly. As a result, the graphics editor was expanded to allow simulation developers to construct switches which the student could set, indicators which came on when appropriate, and relays which changed state when energized.

Figure 10 is an example of the type of display which a developer could produce. In addition to constructing the graphic network, the simulation developer also entered alphanumeric data, via a different editor, which specified the routing of each signal. GAIN used these data to trace signal flows among the switches, blocks, and indicators.

GAIN's signal-tracing capability, which was implemented via a truly trivial subprogram, allowed student and machine to interact about the significance of tests and symptoms in a most natural and intriguing way. The greatest fault of the process lay with the oversimplification required to characterize all cause–effect relationships in terms of simple connectivity. Only in toy systems and ideal systems are failures in a block evidenced at all indicators receiving outputs from

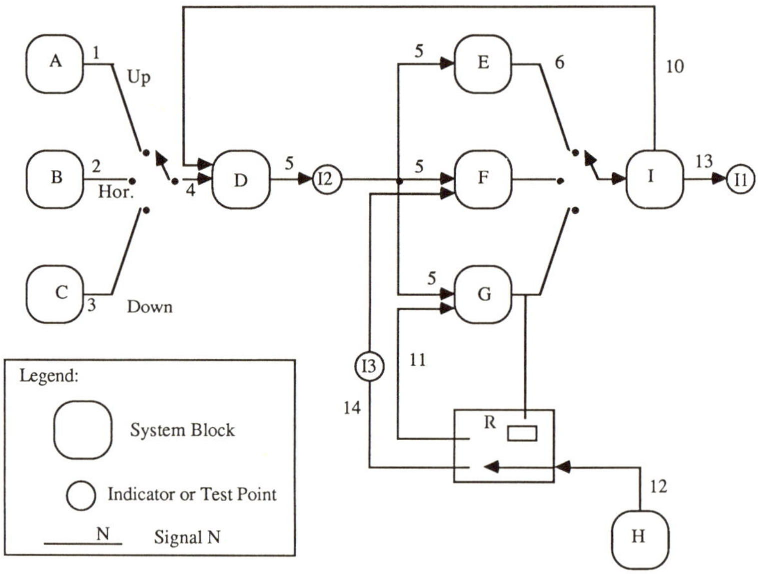

Figure 10. Sample GAIN network.

that block, and only in such systems does a normal indication prove that all blocks supplying signals to it, either directly or indirectly, are operational. In the real world abnormal signals can pass through indicators without a trace, only to be evident again at some later indicator which monitors a different characteristic of the signal, or one which is more sensitive to the abnormality.

The second limitation in GAIN was that the connectivity model did not allow computing the particular symptoms at an indicator. Even if the connectivity assumptions were perfect for a system, the connectivity information only specified whether an indication would be normal or abnormal, not what the symptom would be. Thus GAIN could determine that a meter would read abnormally under certain malfunction conditions and switch settings, but it could not determine the value of that reading. While this was not particularly bothersome at the time, this limitation did prevent the use of the connectivity model in the next generation of generalized training systems. The work in GAIN, however, set the stage for current research in modeling system behaviors, to be discussed later.

For now we were satisfied that GAIN could identify what elements should be suspected, following an abnormal test reading, and which should be exonerated following a normal test reading. By continually updating the set of current hypotheses, using Bayes' theorem, GAIN could compute good next tests, *based upon the student's individual prior testing actions,* and it could evaluate the student's proficiency in selecting his own tests.

D. Branching Out to Performance Aiding

Many who observed TASKTEACH and GAIN in the early 1970s were struck by the fact that the same software and hardware which was serving a training function could just as well operate in a job-aiding mode. Since GAIN was capable of selecting and interpreting diagnostic tests, it should be available to assist in the isolation of actual failures in the field. Further, if GAIN were available to a technician on the job, it could supply additional training and maintain the technician's skills as time allowed.

There were two basic reasons why GAIN could not instantly be cast in this expanded role. First, the computer hardware was still far too bulky and limited to go where the maintenance work was being performed. Secondly, while GAIN could provide a good environment for drill and practice in troubleshooting and procedural skills, it did not deal with many kinds of information which a technician would need on the job. Examples include how to find particular system components, what they look like, and what they do.

As a result we undertook two research and development projects aimed at bringing computer-based intelligence to the technician in the field.

1. MITIPAC-A Portable Maintenance Aid

The first of these was primarily a hardware design and integration effort, which had the objective of producing a portable, fully generalized maintenance training and aiding unit. At the time the notion of a lap computer was just a vision in the minds of people like Alan Kay (1975, p. 15):

> Imagine having your own self-contained knowledge mainpulator in a portable package the size and shape of an ordinary notebook. Suppose it had enough power to outrace your senses of sight and hearing, enough capacity to store for later retrieval thousands of page-equivalents of reference material, poems, letters, recipes, records, drawings, animations, musical scores, waveforms, dynamic simulations, and anything else you would like to remember and change.

Our objective was to produce a unit as much like Kay's Dynabook as possible, as an automatic tutor to the maintenance technician. I termed the system we were seeking MITIPAC, foolishly believing that the name alone would suffice, as Kay's Dynabook had, to convey the general intent of the system. When pressed for what MITIPAC stood for, I managed to come up with Modular Integration of Training Information by a Performance-Aiding Computer.

The specifications for MITIPAC are shown in Table 3; also shown are the successes and failures we experienced for each specification.

The system we actually achieved, shown in Figure 11, fell considerably short of the specifications. While the size and weight goals were not exceeded to a great extent, the operational capabilities of MITIPAC left much to be desired. Since flat-panel technology was highly experimental, and since a conventional CRT would have been impossible to include within the packaging constraints, no

Table 3. MITIPAC Design Specifications and Attained Characteristics

Characteristic	Design Specification	Result
Weight	25 pounds	30 pounds approx.
Volume	1 cubic foot	1.5 cubic feet
Graphic Display	High-res'n color, 160 images	196 color images
User Entry to Graphics	Touch-screeen, 0.5" accuracy	Touch-screen, 0.5" acc'y
Alphanumeric Display	12 line of 40 characters	None
Audio Output	Computer-generated	Pre-recorded on tape
Random Access memory	16K Bytes	16K Bytes
Hardcopy output	Thermal printer	Thermal printer
Mass Storage	Mini-floppy disk or tape	Tape

computer-generated display was provided. Furthermore, since speech synthesis technology was very new, the audio output had to consist of pretaped statements rather than computer-generated responses. As a result of these two shortcomings, MITIPAC was seriously restricted in responding with generated outputs of the type which TASKTEACH and GAIN could yield.

Most importantly, there was only one technology at the time which could hope to meet the color graphics specifications. This was a highly experimental technique which employed a single microfiche plate which contained 196 separate images. The idea behind this approach was novel; each of the 196 images was distributed over a micrographics plate, similarly to the way a hologram is configured. A similarly distributed bundle of fiber-optic fibers transmitted the image a few inches to the viewing screen. By moving the plate a fraction of an inch, a new image came into view. The success of this highly exploratory approach was marginal in MITIPAC. The prototype worked sufficiently well to demonstrate the feasibility of the approach, but the manufacturers had difficulty maintaining the accuracies required to replicate the plates and viewing mechanisms.

While a production version of MITIPAC was never produced, it may have influenced later attempts by others to produce a portable, general-purpose job-

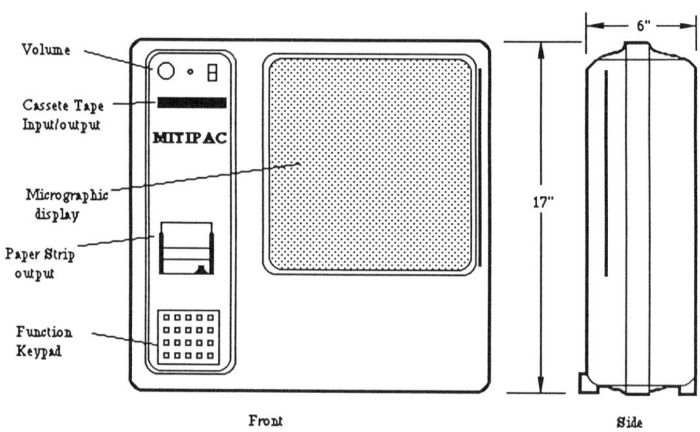

Figure 11. MITIPAC training–aiding unit.

aiding and training system. More than ever we could see that the limitations were not in computer processing, input–output, or memory capabilities, nor in associated software. The shortcomings of MITIPAC were a result of limitations in media for student–computer interactions, specifically small alphanumeric displays, compact graphic capabilities, and accurate touch-input devices. These would plague us for many years to come.

Today, of course, many of the MITIPAC subsystems can be purchased off the shelf of our neighborhood computer store. A true Dynabook for maintainers, or anyone else for that matter, however, still does not exist according to the original specifications.

2. *Managing Equipment-specific Data: The AIDE System*

Our other foray into the world of on-the-job training and performance aiding was concerned with extending the generalized concepts of TASKTEACH and GAIN beyond drill and practice. The AIDE (Automated Instruction, Direction and Exercise) system (Towne and Rigney, 1979, Towne, 1979) was developed to also present instruction on the purpose and theory of operation of equipments and their subsystems, to provide direction in the performance of both procedural and diagnostic tasks, and to serve general information retrieval functions for the technician attempting to either learn on the job or simply satisfy a pressing maintenance problem.

Table 4 lists training and job-aiding requirements for technicians at four different levels of competence and experience. The philosophy of the AIDE system was that equipment-specific data could be extracted from an equipment expert and stored in a standardized format, yet be delivered to a technician in a wide variety of ways depending upon the job requirements and the individual technician. While the emphasis of AIDE was upon establishing the means for organiz-

Table 4. Requirements for Training and Aiding on the Job

Technician's familiarity with particular equipment	Technician's needs for performing job	Technician's needs for on-job training
NONE	Detailed task specifications with graphic aids for locating and identifying system elements	Basic drills in recognizing and locating system elements
MODERATE Can locate and recognize many system elements	Detailed task specifications with on-demand support in locating and identifying system elements	Introduction to purpose and operation of major system elements, and to the basic physical and functional structure
CONSIDERABLE Knows location, appearance, purpose and operation of major elements	Check lists, condensed procedure listings, etc., with on-demand availability of detailed task specifications	1. Elaboration of purpose, structure, and operation of sub-elements. 2. Exercises in symptom interpretation
EXTENSIVE	Easy system for retrieving check-lists, wire lists, schematics, etc.	Simulated troubleshooting problems for maintaining skills and knowledge

ing such information in a generalized manner, the plan was to ultimately develop a wide range of functions for assisting the technician.

My interest was primarily focused on overcoming an age-old flaw in the concept of on-the-job (OJT) training: inexperienced technicians are usually kept away from functioning systems, as their exploration can lead to unnecessary failures and injuries, but when their efforts are required to assist in restoring a failed system there is usually no time to include instruction as well. As a result, the doses of instruction received during job experience tend to be very diluted and haphazard. An automated *and personally tended* system could be advised by the using technician whether or not a particular maintenance requirement must be resolved in the shortest possible time. If so, the system could filter out any unnecessary interactions between itself and the technician, with the purpose of resolving the maintenance problem as quickly and safely as possible. Later, the system could debrief the technician by ''replaying'' the maintenance episode and could provide detailed explanations and evaluations of outcomes via further interaction with the technician.

All of this, however, would require access to a wealth of equipment-specific information, stored in a standardized format. As with earlier systems, the developer of an AIDE data base was not burdened with designing instructional interactions or with planning pedagogical approaches or sequences. Nor did the developer have to specify domain-specific diagnostic strategies, as these were computed in a generalized manner. The domain expert using AIDE supplied lists of control names, indicator names, and failure effects. In addition, the developer could enter information about the location, purpose, and theory of operation of each subsection of the equipment and could key these descriptions to photographic images he supplied.

While impressive (monochrome) computer graphics capabilities were then emerging, their creation entailed exceptionally laborious efforts, and their display required very costly hardware. As a result we continued to rely upon 35mm color slides as the graphic medium. The great advantages of this choice at that time were that very-high-quality photographs and line drawings could be economically integrated into the AIDE data base, they could be easily modified when necessary, whether due to equipment changes or outright errors, and they could be presented reliably and responsively under program control. The obvious limitations were that the number of possible images was highly limited (130 images under computer control), and the images could not respond to user actions.

There was nothing very remarkable about the ways in which domain-specific data were stored in AIDE. The programs which interacted with the equipment expert simply provided a standardized format for the entry and storage of data records specifying a wide range of information types. The expert was prompted to name all the equipment front-panel controls, all the indicators, test points, and adjustment elements; he was guided in breaking a complex system down into a manageable hierarchy and then given the opportunity (but not required) to enter textual explanations of subsystem purposes, functions, and theories of operation. Clearly these latter information items were not processed in powerful ways by the instructional system, and one might object that AIDE was simply storing in computer much of what could be found in a technical manual. While this view is essentially correct, the point of AIDE was to capture in one centralized data base all the information which one might require and to manage that data base in a manner which would facilitate the rational delivery of information under a wide range of situational demands.

The hardware configured to support AIDE bore an interesting relationship to the MITIPAC work. Since the MITIPAC project revealed the critical compromises which were necessitated by requiring portability, we configured the AIDE workstation without regard for size or weight with the idea that miniaturized subsystems would emerge gradually in the future. To get on with the work of developing powerful instructional software, we configured a functional equivalent of MITIPAC which turned out to be approximately 100 times as large.

Constructed entirely of off-the-shelf subsystems, the AIDE station offered a standard alphanumeric CRT; a rear-projected micrographics screen capable of displaying any of 1800 color microfiche images within 3 sec; touch input via a sonic pen which could sense its position on either screen within 0.05 in. on a good day; a hard-copy printer; and a Votrax voice-synthesis system, all controlled by a standard microprocessor. This station was the first for us which approached satisfactory bandwidth of information flow to support powerful instructional and aiding interactions.

Based on this work, we drew several conclusions which influence our work today. First, it is clear that a generalized format can be used to manage information about specialized equipments, not just for instruction in diagnosis, but also for efficient retrieval of detailed technical data and for assisting in the performance of maintenance tasks. Second, we concluded that formats for storing equipment information could survive relatively profound changes in the functions which would convey the information to the user. In other words, we may continue to revise and expand the instructional functions, based upon further research, without having to scrap the representations of the specific systems. In essence, it is feasible to regard a system like AIDE as an open-ended framework upon which we and others could build a wide assortment of specialized functions.

Having access to even moderately powerful processes for interacting with the student quickly revealed the next major deficiency in instructional software: our need and inability to simulate behaviors of the real equipment in response to student actions. TASKTEACH had allowed some student actions, but only in a restricted space of predefined equipment modes; as long as the student set the switches to match a recognized mode, TASKTEACH could refer to the symptom-malfunction matrix corresponding to that mode to determine how the real equipment would behave. GAIN allowed relatively free switch setting actions by the student technician, but its generalized model of system behavior was limited to tracing connectivity. While it could trace a signal from indicator back to a source, it was incapable of determining how an indicator would appear under normal or abnormal conditions, except in the special situation where an indicator had only two states, abnormal and normal. Thus we set out to develop one more, new and improved, trainer, one which would simulate the real equipment as the student technician operated it.

E. GMTS: A Simulation-based Maintenance Trainer

For the first time, sufficiently rich and controllable media existed to consider the development of a simulation-based trainer. The 1800-image color microfiche system used in AIDE was sufficient to simulate the (steady-state) behaviors of a relatively complex real equipment. More accurately stated, the hardware now existed to portray the behaviors of real equipments if computer software could be developed which could compute the correct equipment responses to student actions and to failed components in a generalized manner.[1]

1. Design Options

We considered three general approaches to the problem of providing simulation capabilities within a maintenance training system. If no way could be found to produce generalizable simulations, we would abandon the generalized approach and build in specialized simulation programming for each equipment. This would still offer some benefits over conventional training; moreover, the training functions might be generalized, even in a specialized simulation environment.

The second alternative was to limit the actions of the student, as was done in TASKTEACH, so that we could know in advance what limited states the simulated equipment would enter. This would allow the administration of mode setup drills, in which the student selects all the switch settings required to attain some prescribed mode. The training system would respond by projecting an image of the equipment in that correct mode and go on to the next practice mode.

This alternative would also allow the presentation of diagnostic exercises in which the trainer displayed an image of an equipment front panel for the student to assess. The student would examine the switch settings and the indicator readings and judge whether or not the readings were normal. If they were not, the student could be exercised in interpreting the symptoms in terms of possible failure states. This could also be extended to include evaluation of test equipment readings.

This alternative meant that students could not be allowed to work in a natural way, performing tests freely without constraint. It meant that the student would have to identify the mode which he would set up next by finding the mode name in a list of some sort. Finally, if this option were taken, the displays of all the equipment indicators would be invalid from the time the student started setting up a mode until all switches were in their "proper" position. The environment this alternative would create was considered to be too artificial for further consideration.

The third alternative was to somehow specify what the observable behaviors of a real equipment are in a manner which would avoid having to also represent the underlying functional causes of those symptoms. I am tempted to claim that we carefully considered the possibility of modeling the system behavior in some deep, functional, and generalized fashion and rejected the notion for some well-understood technical reasons. In actuality, the possibility of simulating a system as complex as the AN/URC-32 with a deep *and generalizable* model was then not even formulated sufficiently to state why it could not be done.

The first compromise necessary was to regard even very complex systems as existing in a finite number of states. This meant treating continuous controls in some artificially discrete fashion, such as "fully counterclockwise," "low," "midscale," "high," and "fully clockwise." While this seemed entirely palatable in terms of retaining considerable realism, the simplification only reduced the number of system states for most equipments from infinite to a very large number. The AN/URC-32 transmitter–receiver, one of the target equipments implemented as TASKTEACH and GAIN data bases, could be put into 12,390,000,000 different states if the continuous controls for volume and power were limited to just five settings each. Of course, this huge number is composed largely of meaningless states. Fully one-half of them are states with the main power switch set to off; thus all the variations in the remaining switches are irrelevant—each indicator shows only one reading for all these 6 billion system states. The problem remained to find a way to characterize such behavior. The challenge was considerable, but it was also unavoidable if we were to retain the generalized character of TASKTEACH.

The approach described below dealt first with specifying how a fault-free equipment behaved; failure effects were then expressed as exceptions to these effects.

2. *Specifying Behaviors of Fault-free Equipments*

The technique I worked out was to specify, without enumeration, how each system indicator behaved in terms of the settings of switches which affected it. The following provides the flavor of this approach; while the logical meaning of the expressions given below are accurate, the actual syntax used was considerably more sparse, as computer memory was still a highly limited commodity.

The setting combinations were specified in Boolean expressions, each of which could describe a large set of specific switch setting combinations. For example, condition 1 shown below describes 12 combinations of switch settings:

1: Switch 1 in Position 1, 2, or 3 AND Switch 2 in Position 3, 5, 7, or 9

Each condition like this, which contained only references to switches, was termed a *basic condition*. Compound conditions could then be constructed of basic conditions, as exemplified here by condition 15:

15: Condition 1 OR (Condition 2 AND Condition 3)

This says that condition 15 is TRUE if either condition 1 is TRUE or both condition 2 and condition 3 are TRUE. In this manner conditions can be constructed which specify huge numbers of particular switch-setting combinations.

One could then construct a predicate which specifies the state of a particular indicator, here indicator 17, as

Indicator 17:
If Condition 5 then State 3.

This says that if condition 5 is true, then indicator 17 will be in state 3. The possible states of each indicator are listed separately in the data base, so that the state number can be interpreted in terms of a real-world value when necessary, such as ON, OFF, or BLINKNG. The complete behavior of a particular indicator can be specified by listing all the condition–state pairs required to define the possible states of the indicator. Thus the complete behavior of indicator 17 might be specified as

Indicator 17:
If Condition 8 then State 3
If Condition 22 then State 1
If Condition 5 then State 2

Great economy of specification and evaluation was obtained by evaluating the condition–state pairs in the order listed and stopping evaluation when a condition in a pair was found to be true. This allowed the domain expert to specify common cases high on the list, thereby shortening the compute time. Furthermore, the list could terminate with a condition which was always true, so that the final state represented a default state of the indicator if none of the pairs was true. For example, condition 0 was always defined to be always true, so that the terminating pair was always

If Condition 0 then <some state>

This mechanism was especially powerful when an indicator was only in a state such as ON under special conditions. Then the list of condition–state pairs could list those conditions, ending with the default of OFF.

2. Specifying Equipment Behaviors in Failed Conditions

The foregoing technique was sufficient to specify how each indicator behaved *when the system was functioning normally*. It remained to specify how malfunctions of interest (to be simulated) would alter normal behavior. Equipment behaviors in the presence of a particular malfunction were specified by listing only the ways in which the malfunction caused abnormal indicator behavior. If a particular malfunction only affects indicators 3 and 8, for example, only those two indicators would be specified for the malfunction. Furthermore, only the abnormal behaviors of the two indicators would be listed. Thus, if malfunction 19 only affects indicators 3 and 8 in minor ways, its entire specification might be as follows:

Malfunction 19:
  Indicator 3:
    If Condition 7 then State 3
    If Condition 22 then State 1
  Indicator 8:
    If Condition 5 then State 2
    If Condition 4 then State 1
    If Condition 0 then State 3

By specifying only exceptions from normal behavior in this way, considerable representational efficiency was achieved.

Based on the techniques described above, the Generalized Maintenance Training System (GMTS) was developed. A more appropriate name might have been CHEAT or PROMPT, as the simulation resulted not from an expression of function but of superficial behaviors. The obvious advantage of the approach was that it worked; no other simulation system could take on the complex behaviors of a wide range of systems as could GMTS. This did not come without some expense, of course. To load the GMTS data base with all the correct behavior expressions for each indicator was a considerable chore for a content expert. It involved some person-months of effort to specify how a normally functioning equipment behaved, plus additional hours or days to add the abnormal effects of each malfunction to be simulated. Specifying faulty behavior was especially difficult, as it required the content expert to either theorize all the possible effects of a fault or to actually insert the fault in a real equipment and observe the effects. Both approaches involved considerable time, expense, and skill.

The production of graphic images was also not a trivial proposition. Not even a videodisc (which was not yet on the market) could have stored all the possible front-panel appearances of a typical equipment in complete images; therefore, it was necessary to partition the front panel into subsections. If it was found that a subsection yielded too many such combinations, it was broken into smaller parts yet, until sections were reached which involved a manageable quantity of appearances. A typical equipment subsection required between 8 and 50 slides, each of which showed exactly the same equipment elements, but in different appearances. Color slides were then shot of each subsection, in all its possible combinations of switch settings and indicator readings; color microfiche were then produced, containing 80 images each.

For training effectiveness the motivation was to work with the largest equipment sections possible, for the student would then be seeing more of the equipment at one time. When greater levels of partitioning were required, the student received a more restricted view of the equipment at any one time and was forced to interact more heavily to survey the entire equipment. Ease and economy of media development argued for limiting the equipment subsections, to keep the number of photographic images within reason. At the limit, of course, one could have photographed each single equipment element in each of its possible appearances. This would have greatly limited the number of images and would have simplified encoding the microfiche for retrieval under program control. The student would have suffered great agony in this case by being forced to view the simulated equipment through a veritable keyhole.

The foregoing might leave the incorrect impression that GMTS could only deal with front panels. While I have used this terminology to indicate that GMTS dealt with observable hardware, there was no limitation on the parts of an equipment which could be simulated. GMTS simulations typically included test point strips and even some internal equipment hardware. The point is that GMTS was based on a physical representation of the real equipment. This greatly pleased those who wanted access to high-fidelity substitutes for the real equipment for training purposes, and it bothered those who correctly observed that "GMTS doesn't teach you, it just simulates."

Data bases for two different Navy systems (a Fleet Communications System and the AN/SPA-66 Radar Repeater) were produced, and GMTS training was given to 30 Navy students, thereby demonstrating the generalizability of the technique (Rigney et al., 1978, 1978). The second equipment implemented was specified entirely by two Navy instructors who had no programming experience or previous experience with our representation systems. This demonstrated that content experts could transfer their knowledge of a complex system into a data base for processing by a simulation-based training program.

In 1981 GMTS was given its most rigorous test by placing it in a conventional Navy technical training school and measuring its ability to serve in place of actual equipment for troubleshooting practice (Towne, Munro, and Johnson, 1983). In the first experiment of this study a control group of 31 students received standard training on the AN/WSC-3 satellite communications system; i.e., they attended technical lectures and then individually practiced fault diagnosis by troubleshooting 10 faults inserted into the actual equipment by the instructor. An experimental group of 30 students attended the same lectures, used GMTS to practice eight simulated malfunctions, and then practiced the

Table 5. Experimental Results: GMTS Training vs. Conventional Training

Measure	Experiment 1		Experiment 2	
	Exper. Group	Control Group	Exper. Group	Control Group
Practice Problems:				
Time required(Min.)	50.3	50.1	48.2	46.8
No. Replacements	9.5	9.7	10.3	10.6
Actual Test Problems:				
Time required (Min.)	59.3	59.7	50.5	52.3
No. Replacements	11.2	10.9	8.6	8.7

Note: None of the differences is significant.

remaining two faults on the real AN/WSC-3 system. The control and experimental groups were made as alike as possible by randomly assigning matched pairs of students to the groups. Following the training all students took the standard troubleshooting performance test *on the actual equipment,* the AN/WSC-3 satellite communications system.

There was no statistically significant difference between the two groups, either in time required to solve the faults in the actual equipment or in number of incorrect replacements made (see Table 5, Experiment 1). The control group (conventional training) indicated by their questionnaire responses, however, that they felt their training was more enjoyable and more effective than did those using GMTS.

In a second experiment the experimental group used GMTS for *all 10* practice troubleshooting problems. Again, the differences between the groups in troubleshooting performance on actual equipment were not significantly different. Again, the control group rated their learning as more enjoyable and effective than the group using GMTS.

One aspect of the experimental conditions worth noting was that the total troubleshooting practice time was held constant for the two groups; thus the study compared training effectiveness rates for the simulator vs. the actual equipment and did not allow the greater availability of the simulator to influence the results. Unfortunately, it was not possible to also compare the final competence levels of two groups, one of which had free use of GMTS. While the instructors and training administrators went to great lengths to cooperate and support the studies, they did not feel they could offer one group of students more training than another group. Even for the study which was conducted, a necessary condition was that the control group receive training on GMTS after the study was concluded.

3. *Generalizability of Training Equipment*

GMTS was designed with one further goal in mind: to make the instructional software as independent as possible from the hardware which would execute it, so that GMTS delivery systems could be continually upgraded using off-the-shelf hardware units. The selection of UCSD Pascal for the programming language was largely based upon its ease of portability to a wide range of stand-alone microprocessors. In addition, the Pascal driver routines for the various peripherals were written in a manner which allowed for easy replacement when improved input–output technologies became available, and standard RS-232 interfaces were used between peripherals and the central processor.

One obvious upgrade which we anticipated was the substitution of videodisc for microfiche display. This change was made when videodisc technology matured and when the GMTS system was produced commercially. Unfortunately, the concept of utilizing only off-the-shelf systems for GMTS was lost at this time, since the commercial version included some specially produced hardware. Thus the ability of the users to easily upgrade their systems was lost, and the training hardware could only be acquired from a single producer.

The commercial version of GMTS, called EEMT (Electronic Equipment Maintenance Trainer) did, however, continue to employ the standard GMTS software. The system is now an official training device for the U.S. Navy and is being used to train maintenance technicians on a growing number of equipments.

F. Combining Some GAIN and GMTS Features: The ESAS[2] System

During the early 1980s great advances had occurred in computer graphics capabilities. High-resolution graphic displays were now available on very moderately priced systems. Accompanying these hardware improvements were higher-level software functions which provided the system developer powerful tools for creating and manipulating graphic displays. Our earlier work with the GAIN system had demonstrated the potential of computer graphics; now the opportunity to realize those benefits was at hand. Under a project funded by Lockheed California Company, Allen Munro and I designed a new system which combined the best of GAIN and GMTS. The resulting system, named ESAS (Equipment Simulation Authoring System), used the indicator state rule formulation of GMTS to compute the states of indicators but used computer graphics for the simulation. As a result, the laborious task of filming thousands of front-panel setups and producing microfiche (GMTS) or videodisc (EEMT) media at considerable cost and delay was replaced by the more agreeable and controllable task of constructing graphic representations of individual front-panel elements. The simulation-authoring environment which emerged bore little resemblance to the experimental tools developed for GAIN. The high-level system functions made possible the creation of a graphic editor which allowed for easy creation, editing, copying, and storing of highly realistic front-panel elements. This in turn provided a way to partition the simulation-authoring task into two tiers, one involving the creation of new graphic elements, the second requiring only the retrieval and placement of existing elements into new simulations. Thus two graphic-editing environments were created, one whose purpose was to expand a library of generally useful graphic objects, and the other whose purpose was to construct equipment-specific simulations using the existing library elements. By the time the first few simulations were created in ESAS, the library of graphic elements was sufficiently extensive that new systems could be configured almost entirely of existing graphics.

1. The ESAS Simulation-Authoring Environment

The impact upon the simulation-authoring process was profound. Unlike the GMTS user, who had to wait some weeks or months for the microfiche or videodisc to arrive, the ESAS author assembled the simulation on the screen, keyed in the underlying indicator behavior rules, and then ran the simulation. If an indicator did not behave correctly, the author could examine the underlying rules for the source of the problem and see the effects of the correction immediately.

A second consequence of the ESAS authoring environment was that complex simulations could be authored incrementally. Whereas the GMTS author had to plan for all the desired effects from the early filming through creation of the microfiche or videodisc, the ESAS author could build a little at a time, checking each part before going on to the next.

2. New Training Options

While the most obvious advances of ESAS were related to simulation authoring, significant implications for training techniques resulted as well. ESAS offers the simulation author the opportunity to create several variations of a simulation, the simplest of which might contain only the primary front-panel elements. A novice could be introduced to a complex system via the simplified version and then progress to more complete versions. This ''part-task'' training approach could be accomplished at virtually no cost to the author, as all that is required is that the first version containing only the key front-panel elements be saved intact before adding in further elements.

A second feature provided in ESAS is finding much heavier use than we originally anticipated. Late in the development of ESAS we added a relatively simple capability which allows an instructor to perform a task on ESAS in a ''demonstration'' mode, causing ESAS to record the instructor's actions on the simulation as well as any directions or technical notes entered via keyboard. When the student plays back the demonstration, prompts such as the following are displayed:

> Now set the calibration control to plus or minus 10 dB.

If the student's performance matches the recorded actions, the simulation responds normally; otherwise the student is prompted to try again. If the student is unable to perform the action correctly, ESAS displays a small hand icon which points to the correct setting and it prompts the student to make the action. Upon performing the actions correctly, the student might observe a note such as the following:

> You should note that the signal reflection meter did not register.
> Now return the system to the receive mode.

This recorded demonstration, which we termed *guided simulation,* provides a visually memorable walk-through of a task, it serves to evaluate a student's performance of a procedural task, and it provides an effective way to remediate the student if he or she is unable to perform the task. The primary limitation of the process is that only one way of doing a complex operation is shown, although this is usually the way real-world demonstrations are made.

This capability forms a third tier in the simulation-authoring hierarchy, providing to the instructor a simple way to add training content which is found necessary at the instructional site. While its original intent was only to provide a means for demonstrating and evaluating performance of procedural tasks, it is also now being used to demonstrate and explain diagnostic approaches as well.

III. CURRENT TRAINING TECHNOLOGY

GMTS, its commercial successor EEMT, and ESAS all provide highly realistic simulation of visible behavior of the actual equipment, which might be termed *surface simulation.* While they do not provide simulation of the internal behaviors of a system, studies previously cited indicate that students can effectively practice maintenance on such stimulators, thereby learning at about the same rate as practicing on real equipment, at a much reduced cost.

These systems, like the Link trainers of the early days of aviation, serve primarily as substitutes for the real equipment, but offer little else. They do not perform functions of an instructor or private tutor during troubleshooting practice.[3] They do not select problems in an adaptive fashion which recognizes an individual student's progress and problems, they do not evaluate a student's diagnostic strategy, and they do not recommend or demonstrate improved approaches. Additionally, because these systems have no model of the true operation of the actual equipment, they cannot generate explanations to the student in terms of functional effects.

These instructional resources are built into a recently completed system termed the Intelligent Maintenance Training System (IMTS). The IMTS performs four major functions:

- It simulates the actual equipment front panel and associated test equipment as students practice fault diagnosis.
- It simulates the functional architecture of the actual equipment as students practice fault diagnosis.

- It acts as a private tutor, to evaluate the student's diagnostic approaches, to assist the student when necessary, and to demonstrate preferred diagnostic techniques on specific problems.
- It acts as a surrogate instructor, selecting problems and evaluating student progress during troubleshooting practice.

Figure 12 presents a view of the IMTS screen during a diagnostic exercise. The top line is a menu with which the student exercises control as follows:

APPEARANCE:	The IMTS displays a graphic representation of the physical object or unit.
AFFECTED BY:	The IMTS highlights all blocks in the functional diagram which are affected by the selected block, in the current switch setting configuration.
AFFECTS:	The IMTS highlights all blocks which the selected block affects, in the current switch setting configuration.
HELP SELECT TEST:	The IMTS calls PROFILE to select the best possible diagnostic action to perform next, in light of the student's previous actions and symptoms received.
OPERATION:	The IMTS returns to operation mode (by responding to student actions) following interactions about the simulation rather than interactions implying manipulation.
REPLACE:	The IMTS program will replace the selected unit for the student as if the student had substituted a known good spare for the suspected element.

The second single line on the screen is used to prompt the student in terms specific to the menu selection made. The next major window displays the functional hierarchy of the system being simulated. In this case the simulation is of a system called Bladefold, a moderately complex, electrically controlled hydraulic system which, after landing, causes the blades of the SH-3 helicopter to brake to a stop, rotate to a designated orientation, and then fold into a compact stowed position. The examples which follow all relate to this particular application of the IMTS.

The majority of the screen is devoted to a more detailed view of one of the blocks selected on the upper window. In Figure 12 the student is seeing the "top level" of the Bladefold system, in block diagram form, and has just selected the Fold Select Valve section for more detailed display. Selections deeper in the functional hierarchy bring up detailed functional or physical views as shown below.

A. Simulation of the Physical World

To an outside observer the graphic simulation generated by IMTS of the actual equipment's physical appearance looks much like that produced by ESAS. A

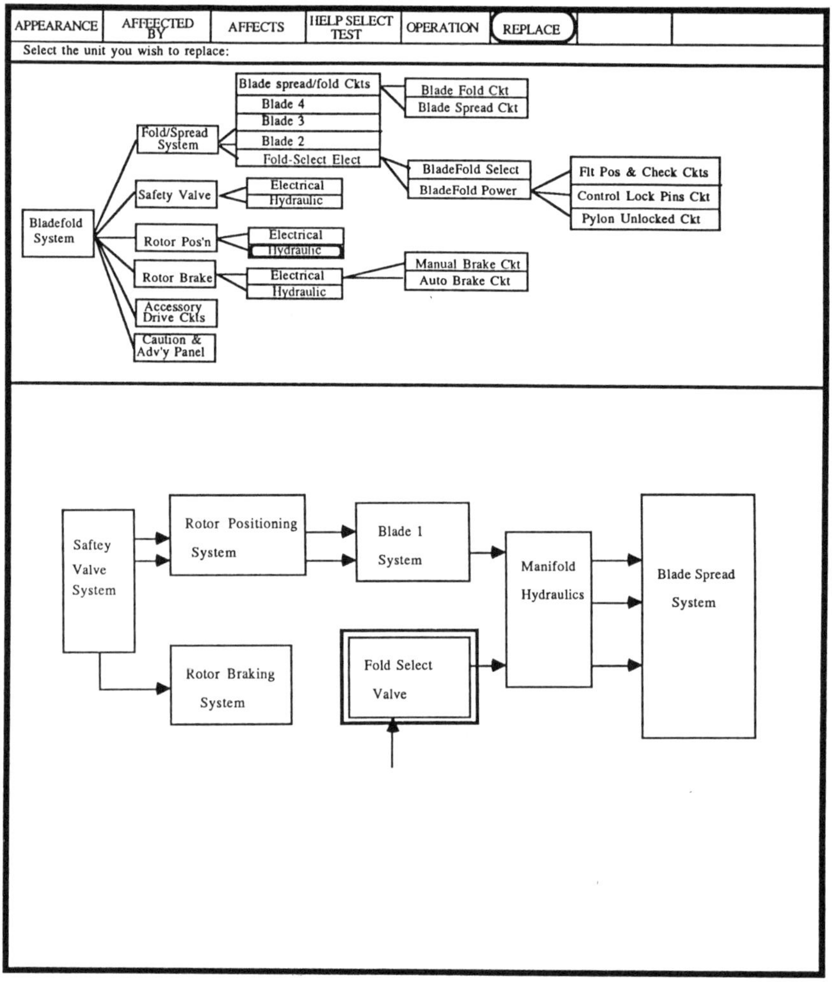

Figure 12. IMTS screen displaying top level of bladefold.

sample of a physical simulation of a portion of the helicopter cockpit is shown in Figure 13.

B. Simulation of the Functional World

Unlike all the training systems discussed previously, the IMTS has the ability to display the internal organization of a system and to simulate the *functional* behavior of the system in response to student actions.

The IMTS determines the current states of equipment indicators by processing the deep functional representation of the system provided by the simulation

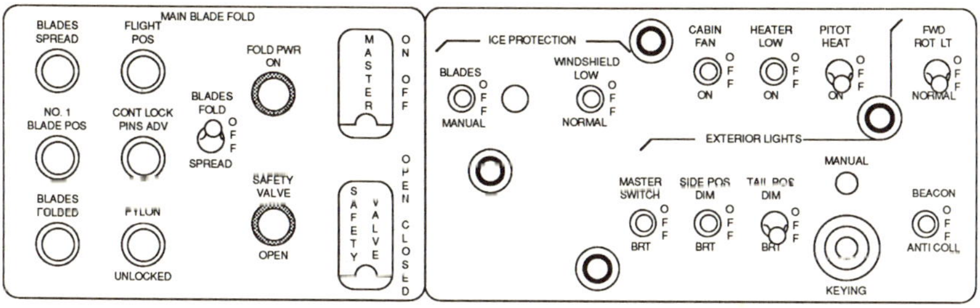

Figure 13. Physical simulation of front-panel section.

author, rather than by applying the numerous and somewhat superficial indicator behavior rules used in GMTS and ESAS. Thus the functional description is central to driving all IMTS simulations, while the surface physical simulation is optional.

When both types of simulation are available for a particular equipment, the student controls the view of the equipment throughout the training session. Since all switches, indicators, and test points are shown in the functional representation and can be operated identically to their physical counterparts, a student may choose to work entirely in the functional mode if he has no need to see the true physical appearance of the system.

Whenever the IMTS notes that the student has changed a switch or altered the (simulated) system configuration in any way, it calls a special simulation program to interrogate each object to determine its response to the change. As each object reports its behavior (the ways in which it processes inputs and the state it enters) it also displays itself. In this way meters, lights, and dials all respond correctly to student actions.

Moreover, other components in the simulated system exhibit their states as well, providing a view of the system that even the actual equipment cannot provide. In this manner students can see the inner workings of systems and can better understand their behaviors.

Figure 14 shows the IMTS screen for the hydraulic subsystem which controls the actuating cylinders. Each object shown behaves according to its own rules; most of them change appearance when they change state. Each object may be failed by the IMTS surrogate instructor program and each may be replaced by the student. In the electrical portions of the system there are numerous switches and cockpit indicators. In Figure 14 there are just a few controls and indicators. Most testing of this subsystem is done by inserting a pressure meter in the hydraulic lines.

There is no way to convey on paper the dynamic quality of the IMTS simulation. A few well-chosen actions on the display shown in Figure 14 cause several dozen of the components to change state; valves open, locks retract, previously inactive pipes come under high pressure, and ultimately actuating cylinders

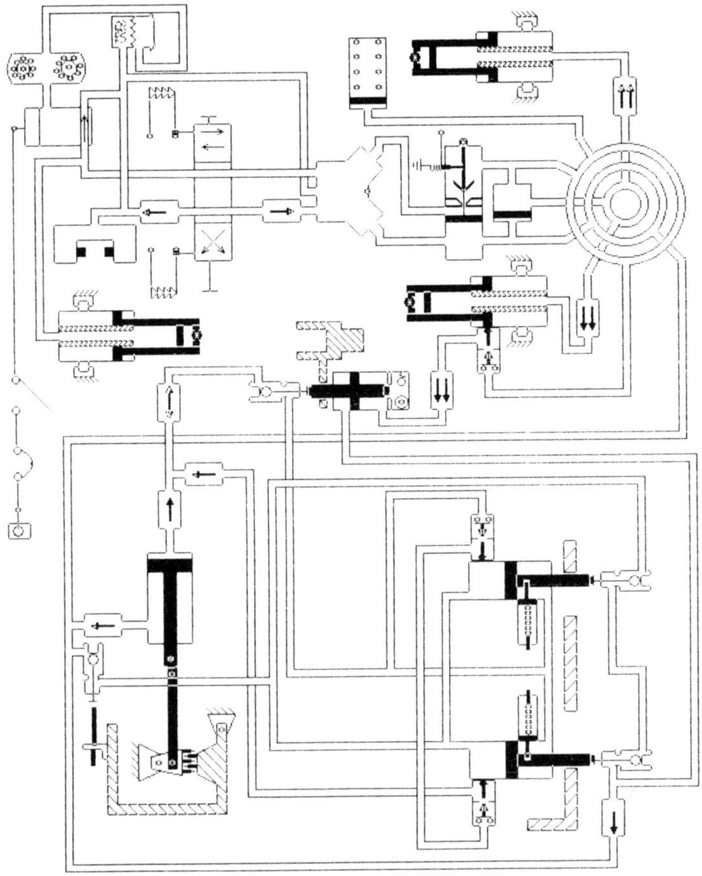

Figure 14. A portion of Bladefold simulation.

extend or retract, thereby folding or extending the helicopter blades. During normal troubleshooting practice the IMTS updates the graphic presentation as rapidly as possible in response to each action by the student. Alternatively, the student may ask to control the updating of displayed responses. In this case each depression of the mouse button causes the simulation to update just one object on the screen so that the student can carefully observe how each object is affecting others and why the sequence of events is as it is.

Soon the IMTS will have the capacity to operate upon multiple representations of a real system *within a single instructional session* (while ESAS allows alternative representations of an equipment, these are independently manipulated during separate instructional phases). One result of this is that the student will be able to refer to more simplified or more complete representations, depending

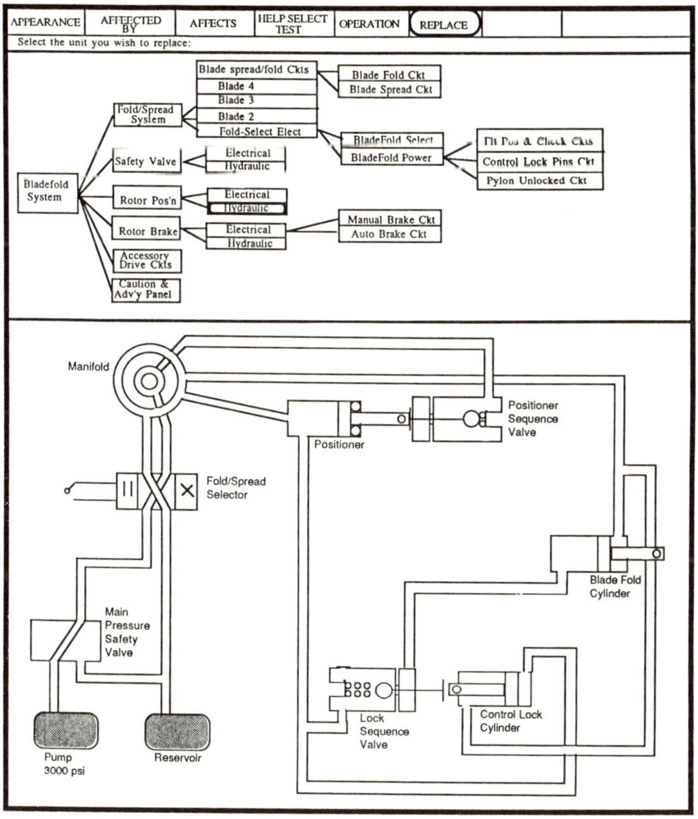

Figure 15. A moderately simplified hydraulic representation.

upon his preferences and current learning situation. Figure 15 presents a somewhat simplified form of the hydraulic subsystem shown in Figure 14, which could be used for training novice technicians.

Figures 16 and 17 present the most simplified representation possible; the two diagrams are identical except that Figure 16 shows the system with the Bladefold Cylinder retracted, while Figure 17 shows it extended. The graphic transition between Figures 16 and 17 would happen on the IMTS screen in response to the student actuating the Fold/Spread Selector. Interestingly, this highly simplified form conveys the essence of the process performed by the system shown in Figure 14, minus handy peripheral functions which prevent the blades from slamming into position with crushing force and which prevent the blades from folding during flight.

In Figure 16 we see just two controls which the student can actuate, the Fold/Spread Selector and a simple toggle switch next to it. When the setting of

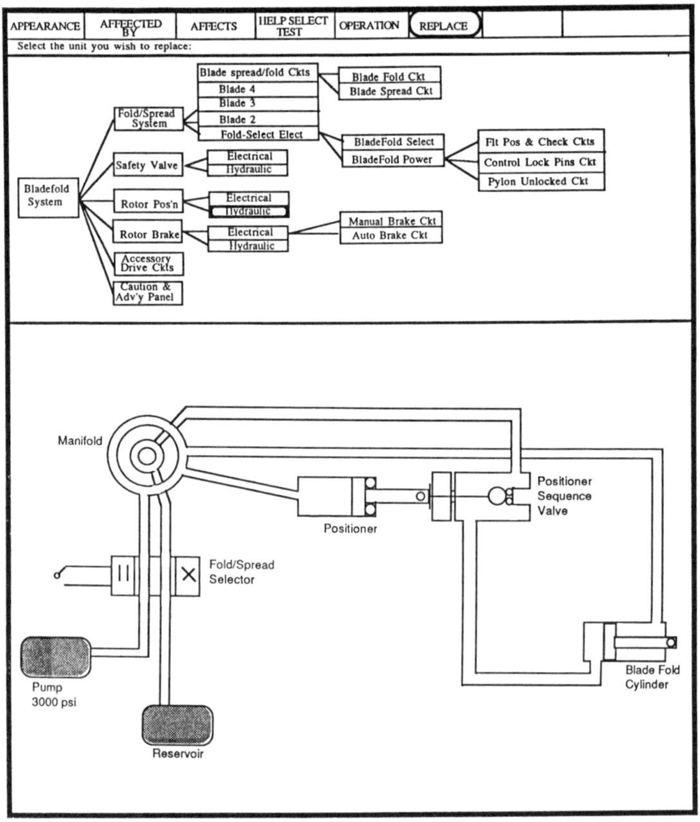

Figure 16. A highly simplified hydraulic representation (Bladefold Cylinder retracted).

either of these is changed by the student, the hydraulic lines from the pump and the reservoir switch to a crossed position within the Fold/Spread Selector, as shown in Figure 17, and the remaining three elements in the system change position. Figure 18 shows the IMTS screen when an electrical subsystem has been selected for detailed viewing and operation.

As discussed below, these simulations are produced in quite a different manner than was done previously.

C. Tutoring Functions

As with GMTS and ESAS, the student may freely perform tests (and ''non-tests'' resulting from errors) and effect the ''replacement'' of components suspected as being faulty. As a result of the functional basis upon which the simulation is generated, however, the IMTS can evaluate a student's diagnostic

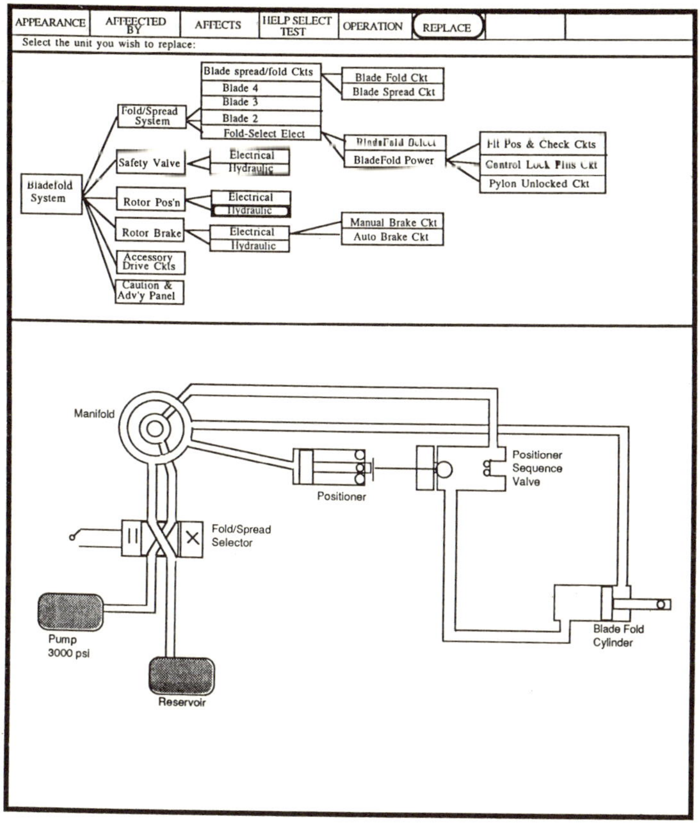

Figure 17. A highly simplified hydraulic representation (Bladefold Cylinder extended).

strategy, it can explain what any test results indicate, and it can demonstrate preferred diagnostic approaches.

To do this the IMTS incorporates a generic model of an expert troubleshooter, capable of operating upon any specific system specification (the functional representation discussed above) to generate an efficient and explainable sequence of testing actions resulting in the identification of a fault. This expert model, termed PROFILE, is the result of several years' research into the ways in which human maintenance experts differ from theoretically optimal behavior (Towne, Johnson, and Corwin, 1982, 1983; Towne, 1982, 1984). Stated very concisely, we found that human experts can be modeled very well by a near-optimal diagnostic process which operates upon very diluted, or fuzzy, system information. Stated another way, expert troubleshooters seem to employ excellent strategies based on very imprecise symptom knowledge. Thus PROFILE performs a search for good

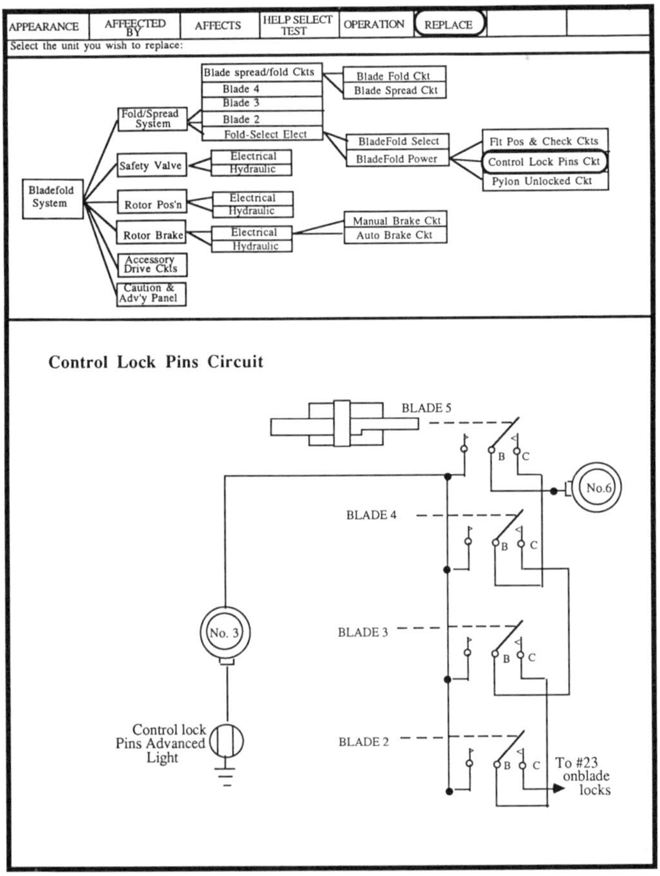

Figure 18. IMTS screen displaying an electrical subsystem.

tests using a near-optimal selection rule based on information theory, but it only uses clear and compelling symptom information to represent its understanding of the system. Accordingly, students find that PROFILE's decisions and justifications for its testing choices are reasonable and not something only a computer could have figured out.

By imbedding PROFILE in IMTS, we provide a resource which can evaluate the quality of each student's diagnostic work, interrupt performance when serious problems are detected, and offer specific guidance for continuing to solution. As distinguished from a classical expert system, PROFILE does not require tedious front-loading of a human expert's diagnostic rules. The expert system approach shown in Figure 19 involves computer software which traces through the rule base to infer facts and relationships not explicitly represented. This approach may yield highly system-specific responses to student actions, but it

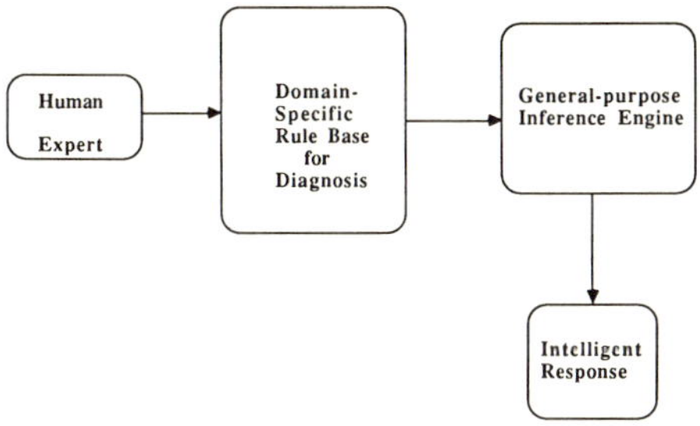

Figure 19. Expert system for incompletely understood system.

does so at great cost of data development. The great advantage of the expert system approach, not required here, is that it will support training the diagnosis of an incompletely understood system, such as medical diagnosis.

The approach employed in the IMTS, shown in Figure 20, is to require the content expert to provide only a technically correct representation of the system, in the form of diagrams. Once entered to the IMTS, via special configuration editor programs, *these diagrams support all the functions of the training,* including simulations which respond to student actions, the evaluation of student diagnosis and operation, problem selection, and generation of expert diagnostic strategies.

D. Surrogate Instructor Functions

Managing each individual student's instruction is a set of functions which serves as a private instructor. These functions select problems for the student according to demonstrated progress and difficulties, they effect simulated replacements when requested by the student, and they call in the expert when appropriate. Problem selection is guided by a model of the individual student which the surrogate instructor constructs as the student works. This model reflects the domain information to which the student has been exposed, and has mastered, according to the problems he has solved. As with the PROFILE model, the functional model of the actual equipment provides the structure of domain information which is involved. Terminal nodes in the system hierarchy

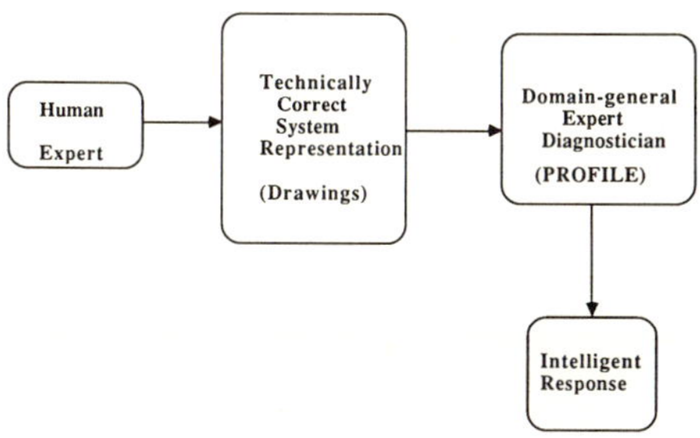

Figure 20. Expert system for completely understood system.

represent individual components which can fail; each of these is assigned a subjective level of difficulty by the content expert. When a student solves a particular malfunction, the IMTS marks the node in the system hierarchy which was experienced.

Selection of the next problem is then computed as a function of (1) the perceived difficulty which is appropriate for the student and (2) the conceptual "distance" between the previous problem and the next problem. Conceptual distance is computed between the just-solved node and all other nodes which offer the appropriate difficulty. This is a function of the node's place in the system hierarchy; the closer a node under consideration is to the just-solved node, the smaller is the conceptual distance.

Instructional theories do not presently provide clear guidelines for setting the appropriate difficulty and conceptual distance for a practice problem. We will undoubtedly experiment with alternate approaches until the problem selection function operates in an acceptable manner.

E. Authoring Simulations in the IMTS

The functional model used in the IMTS was heavily influenced and inspired by work on a simulation system called STEAMER (Hollan, Hutchins, and Weitzman, 1984; Hollan, 1983). STEAMER allows experts to construct simulations of particular systems by constructing "objects," each of which has one or more graphic appearances, plus underlying rules governing its behavior. When assembled into a particular system by a content expert, the objects determine how they behave and how they appear under any condition by executing their own behavior rules. Thus, as a student alters the system configuration, by setting switches, the intelligent objects respond appropriately.

We have developed general-purpose authoring tools which allow an expert to create new object types, which we term *generic* objects, and to construct equipment-specific simulations.

1. Creating a New Graphic Object

If the simulation author determines that the existing library of generic objects lacks a required object, he constructs it using a graphic object editor. This involves constructing on the screen that part of the object which does not change, called the static part, and then entering the graphics which change according to the state of the object. Frequently the graphics for one state of an object can be created by simply copying and moving the graphics for another state. Figure 21 shows an object in its two states.

2. Getting Objects to Behave

Once a new object is defined graphically it must be provided its rules of behavior. This is done within a special generic object behavior editor, as shown in Figure 22. This editor provides four windows as follows:

1. The *Defined Objects window,* in the upper right-hand part of the screen, lists the names of all objects whose graphics have already been created; the user scrolls through this window and selects the name of the object previously created in the graphics editor.
2. The *Object Display window* displays whatever object has been selected; as the user steps through each object state the display cycles to the proper state.
3. The *Constant Object Specification window,* on the left of the screen, prompts the user for information about the object which is not state-dependent. Via menus which come and go as required, the user identifies the types of inputs and outputs the object processes, such as hydraulic vs. electrical or mechanical (the alphabetic labels shown in Figure 22 are only seen by the object creator). Here the user also enters information about the object which is important to the PROFILE troubleshooting model. The replacement time, spares cost, and mean time between failures (MTBF) all influence decisions by PROFILE.
4. The *State Definition window,* at the lower center of the screen, prompts the user for information about each state of the object. Here are entered the rules governing what processes the object performs upon its inputs and what causes the object to enter each state.

The object shown in Figure 22, called a BladeLockCylinder, has only two graphical states, but it has four different *behavior* states. The first two states, Locked and Unlocked, are normal states. Two other possible states are failure conditions: Stuck-locked and Stuck-unlocked.

Virtually all the components of the Bladefold system are elements which exist in a few discrete states, such as extended–retracted, locked–unlocked, or on–off. While these happen to be common in the Bladefold system, the IMTS is not limited to simulating discrete-state elements. The pressure meter, for example, is an object which performs the simple function of sensing the pressure in a pipe and reflecting that reading via a needle. One could just as easily define an amplifier whose function is to output the square of its input. Furthermore, there

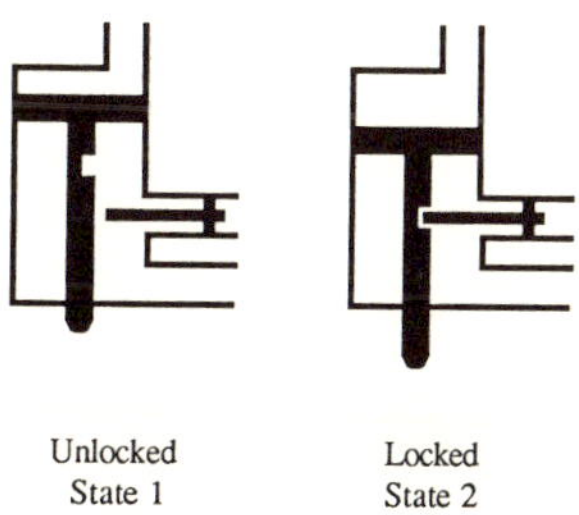

Figure 21. A two-state generic object.

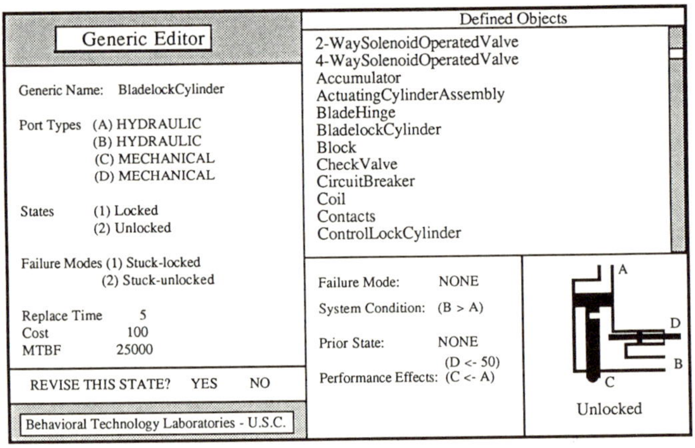

Figure 22. The generic object behavior editor.

happen to be no objects in the Bladefold system whose current state depends in part upon a prior state, such as flip-flop. The generic object editor offers the object definer the means for including an object's previous state into its definition of current state.

3. *The Library of Generic Objects*

When an object has been defined both graphically and behaviorally, it passes into a general library which can be used as a resource by any simulation author. At the time of this writing we have defined only those objects required to simulate the Bladefold system, such as wires, switches, indicator lights, meters, valves, relays, and pipes, yet just these objects would take authors a long way toward simulating many new systems. A portion of the generic library is shown in Figure 23.

4. *Constructing New Simulations for Training*

The content expert constructs a specific system simulation (and all associated training interactions) by simply selecting appropriate objects from the library and positioning them on the screen, using a special graphics editor. Fortunately, perhaps, this is the easiest authoring function. While the author must be certain that each object selected actually operates as intended, the job of constructing the simulation is primarily one of subdividing a big system into separate screens or drawings and then producing each individual diagram in the editor provided with the IMTS. Normally, existing technical diagrams serve as an excellent starting point for creating the diagrams. It is also common, however, that drawings from technical references are (1) wrong and (2) not created for ease of understanding. Typically the simulation author will massage the drawings considerably before being satisfied with the clarity of the displays.

As the objects are positioned, the editor detects the connections between

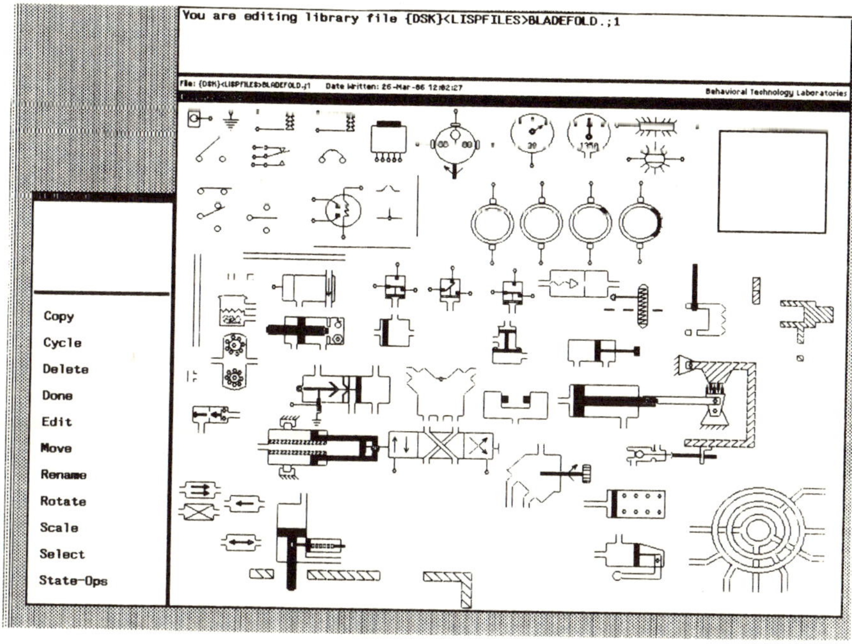

Figure 23. A portion of IMTS generic library.

elements, and it retains the connectivity data in a file. While the connectivity data are necessary to computing how a system will behave under a current condition, these data are a small part of the intelligence used to simulate the system behaviors. Whereas GAIN relied entirely upon connectivity data to compute the *flow* of signals throughout a system, the IMTS uses this information plus the behavior rules of each object involved to determine the *nature of the signal conversions,* and hence the particular appearance of system indicators and associated test equipment.

Once all the individual screens have been created and outputs from one diagram have been linked to inputs in other diagrams, the simulation is completed. IMTS can now select and insert practice malfunctions for each student; it can accept and display the results of student-testing actions; it can monitor each learner, providing individualized assistance; and it can demonstrate expert diagnostic strategies as required.

5. *Operational Implications*

To the IMTS simulator there is no difference in principle between normal and abnormal system behavior. While the on-line expert troubleshooting program keeps track of normal and abnormal symptoms received by the student, for the purpose of debriefing the student on the diagnostic episode, the simulator does not pay attention to whether components it encounters in the functional description of the system are normal or abnormal. To introduce a malfunction into a simulation, then, the IMTS instructor program simply replaces one of the normal object specifications with an abnormal description. When the simulator calls upon that failed component, it receives the correct information for how that failed component behaves. Some attractive implications of this are the following:

- Students may request the introduction of failed components, to explore their effects.
- A wide range of failure modes may be specified for a component, allowing very selective simulation of abnormalities.
- Multiple failures may be introduced with virtually no complication in the simulator or in the authoring process.
- The student or the surrogate instructor program may ask for the simulation of virtually any type of short circuit or open circuit without requiring that the simulation author anticipate the possible requests.
- "Cascading" failure effects, in which a failure in one component causes a failure in a second component, and so on, are simulated correctly, without anticipation by the simulation author.

To incorporate the capability of dealing with cascading failures does require the author of generic objects to describe the conditions under which each object will fail, and to specify which failure mode it enters under each condition. While this can involve some effort, it is a one-time endeavor per generic object. If the student happens to replace a component that failed as a result of another failure, prior to resolving the source of the cascading failures, the simulator will discover that the newly replaced component, such as a fuse, must change back to its previous failed state because of the forces acting upon it and will substitute the failed component into the functional representation. As a result the student will observe symptoms identical to those seen prior to the replacement, since the replaced component has failed again. When the true source of the problems is replaced, the symptoms will probably change, but will still be abnormal until all the resultant failures are also resolved.

IV. INSTRUCTIONAL ISSUES

Serious instructional issues were not encountered when maintenance trainers were only simulators; one simply used the simulator in place of the real equipment. Now that training systems are capable of serving as a surrogate instructor, tutor, and coach, a new wave of instructional questions must be addressed. Some of these issues will be listed here, with very brief remarks about alternative views.

A. Knowledge of Results

As maintenance trainers advance into more complex training domains, the questions about how and when to inform a student about his or her work multiply. Most research shows rather clearly that immediate presentation of knowledge of results (feedback) enhances learning of the task at hand. Immediate feedback, however, has the unfortunate effect of depriving the learner of the experience of monitoring his own behavior and of detecting errors. Certainly self-monitoring is one of the important skills involved in corrective maintenance; thus there must be provided some opportunities to work without constant external correction. Our studies of feedback timing, which were conducted in the domain of real-time tasks as opposed to serial-action diagnostic or procedural tasks, indicates a marked advantage in providing the learner with substantial control over the level of intrusiveness of the instructional system (Munro, Fehling, and Towne, 1985; Munro, Cody, and Towne, 1982).

In considering what knowledge of results should be presented to the maintenance student, it is crucial to recognize that fault diagnosis, like computer programming and many other complex endeavors, is extremely error-prone. Even the finest of maintenance experts operate in an "area of imperfection." This is not to say that they frequently commit serious blunders. But it is true, based upon our detailed observation of hundreds of troubleshooting episodes, that experts cannot draw perfect conclusions about each symptom, they cannot know all the possible implications of each test result, nor can they always (or even usually) be expected to perform optimally.

In the real world, experts are not expected to perform optimally. It is recognized that today's complex systems are too involved to be perfectly understood by individual technicians (Bond and Towne, 1979). Even the designers of complex systems often cannot predict how every malfunction will impact system behavior. This forms a strong case for also not expecting or demanding perfect performance from trainees. A training system which did expect perfection would be one which constantly interrupted and constantly exhibited to the learner just how imperfect he is. While this might be an enjoyable and challenging environment for highly qualified personnel, it does not seem like a motivating setting for beginning or intermediate students.

B. Disclosure of System Effects

During truly instructional phases of maintenance training, when concepts and theories of operation are being presented, it is useful to reveal, and even draw the student's attention to, important aspects of the system's behavior. For example, in Bladefold system training with the IMTS, the graphic simulation is extremely useful in depicting the interactions among the various valves, pipes, and actuators. The simulation is able to reveal aspects of the Bladefold system which would be difficult to discern using the real equipment. During troubleshooting practice, however, these graphic cues can be so powerful as to destroy the fault diagnosis exercise; one look at the simulation screen might reveal what has gone wrong.

For this reason, it is crucial within a simulation-based trainer such as the IMTS to have control over the graphic display, so that some effects may be shown during instruction and remediation, but hidden during troubleshooting practice.

C. What to Teach

At least two levels of instruction might be useful in maintenance training: (1) teaching what to do, to accomplish various objectives; and (2) teaching *how to decide* what to do. The latter approach is more commonly taken in the instruction of such topics as geometry, calculus, or even computer programming, where the process of arriving at a solution can be characterized and somewhat formalized. The military services have flirted with formalizing the instruction of fault diagnosis by breaking corrective maintenance requirements into phases such as system state recognition, symptom gathering, fault localization, fault isolation, repair, and system confirmation. These subdivisions may be useful for early instruction to suggest to the student technician that some rigor be attached to the process of finding a fault. The hope is that such rigor will produce a technician who performs adequate testing and reasoning before replacing components.

At a deeper level, it is relatively unclear whether one learns generalized diagnosis skills better from exposure to explicit and generalized concepts or to a series of domain-specific decisions which follow those concepts. In the former case an automated tutor might say, "You should evaluate the current suspected units and find those which are common to a function which can be easily tested . . . ," while in the latter case the direction might be "You should notice that the two frequency modulators are both monitored by the frequency meter, so you should check it next."

A related issue concerns just what fault diagnosis strategies should form the basis of instruction. We have found that experts do not follow a procedure which simply minimizes expected time or number of tests to isolate the fault. Instead, they take into account the cost of spares which may be consumed in the process and the exposure to error. Outside the laboratory they definitely consider danger and discomfort in forming their plans. While the PROFILE model of expert diagnostic performance does take into account the relative power of possible tests, the past reliability of the possible failed components, the cost of spares, and the times to perform various actions, it does not currently consider the various error likelihoods that accompany each alternative testing action. Nor does it currently consider the cognitive workload associated with alternative diagnostic strategies. While we are doing research aimed at broadening our concept of expert performance (Towne, 1985), it is clear that much remains to be done in this area.

D. Degree of Machine Control and Student Control

Some learners seem to be highly self-directed. They know what it is they are trying to do, and they have some plans about their own learning approach. It seems to make sense to put the training system largely at the disposal of such individuals. The system can provide all sorts of interesting on-demand functions to assist them; otherwise, it seems, the system should attempt to minimize interruption and intrusion.

Other learners perform poorly in such environments. They would rather have the system calling the shots. After all, they have their hands full learning new and complex concepts, so why should they also be burdened with the task of managing their own instruction?

While the former type of learner seems to be more highly regarded than the latter, it is not at all clear if different levels of control really are more appropriate for some learners than others and if self-directed learners really do learn more effectively than those preferring higher structure. In any case, the problem remains of distinguishing individual preferences and abilities along this dimension and of tailoring instruction to best suit individual abilities.

E. Degree of Instructor Control

The ESAS and IMTS systems described previously offer the instructor a number of ways to affect the training. These include mechanisms for adding new content, new problems, additional explanatory remarks, and necessary warnings. In addition, the IMTS provides ways for the instructor (1) to revise the assessment of problem difficulty, (2) to modify the level of performance monitoring provided by the trainer, and (3) to refine information about the assumed knowledge of the students.

From the instructor's point of view, such opportunities for impacting the instruction are regarded highly, especially since these inputs are allowed but not required. From the student's point of view, this instructor input is probably very beneficial to the quality of instruction, for the instructor possesses important information about the student population and previous instruction. The drawback of providing instructor control within automated trainers is that consistency and reliability of instruction could suffer. Without instructor control the student knows just what to expect each time the training system is used; with instructor control anything is possible.

In general, our approach within the IMTS is to provide the instructor considerable control over the degree of intrusion which the trainer will impose and over assumptions about student knowledge and problem difficulty.

F. Problem Selection

The technique used within the IMTS for selecting problems was briefly discussed. It was also noted that since the IMTS operates from a functional model of the target system, it is capable of simulating virtually any malfunction of interest. One pertinent question is: Who in the man–computer team is better suited to select the level of difficulty and conceptual distance between problems? While we have attempted to automate all the problem selection processes, it could be that the student is a better judge of the appropriate difficulty level than any generalized program. Similarly, the student might learn more effectively if he could command the trainer to stay in a particular conceptual area a longer or shorter time. The IMTS partially accommodates these concerns by giving the student a control that influences the ideal difficulty level and conceptual area of the next problem.

Now that we have somewhat refined mechanisms for monitoring student proficiency, we also have the means for administering instruction in a very individualized manner. The problem is one of sensing when the student knows best and when he does not.

V. CONCLUSIONS

Early training systems were seriously burdened with clumsy, slow, and unreliable electromechanical subsystems with limited capacity for simulating the real world and for interacting with the learner. Today the hardware exists to provide highly realistic simulations of real equipments and to allow cognitively realistic operations by the trainee. The flexibility, resolution, and speed of producing computer graphic representations is sufficient to support simulations of a wide range of real equipments, and user input devices such as the touch panel and the mouse provide meaningful ways for the learner to participate in the simulation.

New research and development can be approached within the framework of systems like the IMTS. It is now pertinent and feasible to explore means for generating customized simulation displays under computer control to best meet the needs of the individual learner, to investigate alternate viewpoints for rationalizing expert decisions and strategies, and to study how the unique resources of the simulation author, the instructor, and the student can be coordinated in a fashion which will maximize effectiveness of training.

Other major problems now relate to training effectiveness. Research is beginning to reveal how people learn and maintain complex concepts, how they relate new information to old, and how they respond to different instructional environments. The generalized maintenance trainer is likely to become one of the major research tools, for within such a system instructional variables can be controlled precisely, high-resolution performance data may be accumulated, and a relatively wide base of application and users may be obtained.

There is little doubt that future technology will bring about even more exciting approaches to the simulation of real systems. Today's two-dimensional and relatively silent simulations may be replaced by three-dimensional views of immense spatial capacity and clarity, along with realistic sounds, smells, and sensations.

While these prospects are exciting, the major progress in automated training systems will now come from a better understanding and representation of the human machine.

ACKNOWLEDGMENTS

Virtually all the developments described here were made possible by the continued support from the United States government, primarily the research and development programs of the U.S. Navy. Except for the ESAS system, the training systems described were developed under funding from the Office of Naval Research, the Navy Personnel Research and Development Center, the Naval Training Equipment Center, the Bureau of Naval Personnel, or the Advanced Research Projects Agency.

The following attempts to acknowledge the funding agencies, supporting individuals, and key technical contributors for each major development. The special influence of Dr. Joseph W. Rigney in envisioning the possibilities and then working to secure the support and to maintain a productive research environment cannot be adequately expressed.

BETS

Early work in developing and evaluating the BETS model was done under funding from the Personnel and Training Branch, Psychological Sciences Division, Office of Naval Research, Arlington, Va. (Dr. Glenn L. Bryan and Dr. James Regan supported the work); and from the Personnel Research Division, Bureau of Naval Personnel (Dr. Victor Fields, Dr. Martin Wiskoff, Mr. Sidney Friedman, and Captain L. R. Wilder).

Mr. Richard L. Weis and Dr. Nicholas A. Bond were the developers of the quantitative model. Dr. Robert H. Cremer and Dr. Anthony K. Mason contributed in major ways to the development and evaluation of the model.

TASKTEACH

TASKTEACH was developed under funding from the Personnel and Training Branch, Psychological Sciences Division, Office of Naval Research, Arlington, Va. (Dr. Glenn L. Bryan, Dr. Marshall Farr, Dr. Victor Fields, and Dr. Joseph Young), from the Psychological Research Branch, Bureau of Naval Personnel, Washington, D.C. (Mr. Ray Williams and Mr. Earl Hinman), and from the Advanced Research Projects Agency, Washington, D.C. (Dr. Thomas O'Sullivan).

Ms. Carole A. King took over the bulk of programming implementation of later versions.

GAIN

The GAIN system was developed under funding from the Personnel and Training Branch, Psychological Sciences Division, Office of Naval Research, Arlington, Va. (Dr. Marshall Farr and Dr. Joseph Young) and from the Advanced Research Projects Agency, Washington, D.C. (Dr. Thomas O'Sullivan).

Ms. Carole A. King performed the major portion of the programming implementation.

AIDE and MITIPAC

AIDE and MITIPAC were developed under funding from the Naval Training Equipment Center, Orlando, Fla. Dr. William King served as scientific officer. The MITIPAC prototype was produced under subcontract by Logicon, Inc., Torrance, Calif.

GMTS

GMTS was developed under funding from the Personnel and Training Research Program, Psychological Sciences Division, Office of Naval Research, Arlington, Va. (Dr. Marshall J. Farr and Dr. Henry Halff), from the Navy Personnel Research and Development Center, San Diego, Calif. (Dr. James S. McMichael, Mr. Vern Malec, and Dr. James J. Regan), and from the Defense Advanced Research Projects Agency, Washington, D.C. (Dr. Dexter Fletcher).

Ms. Carole A. King of our organization contributed heavily to the GMTS concept, and she accomplished the majority of the programming implementation. Dr. Allen Munro designed the experimental studies of training effectiveness. Mr. Richard A. Mishler and Mr. Patrick J. Moran served as subject matter experts for three implementations of electronic systems to GMTS.

Mr. Ronald E. Renfro of Mathetics Corporation designed the frame-oriented instructional feature of GMTS, which was implemented by Mr. Mark C. Johnson of our organization. EEMT was manufactured by the Cubic Corporation.

ESAS

ESAS was developed under funding from the Lockheed California Company (Mr. Ray McCaslin). Ms. Roz Schwartz of Lockheed performed all the programming implementation.

Dr. Allen Munro of our organization was the primary designer of new ESAS functions.

PROFILE

The PROFILE model of troubleshooting performance was developed under funding from the Engineering Psychology Group, Office of Naval Research, Arlington, Va. (Mr. Gerald S. Malecki and Dr. Martin Tolcott).

Mr. Mark C. Johnson accomplished the programming implementation and contributed in major ways to the PROFILE design and experimental evaluation.

IMTS

The IMTS system is being developed under funding from the Personnel and Training Research Program, Psychological Sciences Division, Office of Naval Research, Arlington, Va. (Dr. Susan Chipman, Dr. Michael Shafto) and from the Navy Personnel Research and Development Center, San Diego, Calif. (Mr. Vern Malec).

Dr. Quentin A. Pizzini was responsible for developing the generic object behavior editor and associated underlying data structures within IMTS, he implemented PROFILE in Lisp, and he produced the debriefing and strategy evaluation capabilities of the IMTS. Mr. David S. Surmon developed the generic object graphic editor, the specific-system simulation editor, and the IMTS simulator processor program. Dr. Allen Munro was responsible for the development of the student modeling and adaptive problem selection processes in IMTS. These three have made major contributions to the design of the IMTS throughout the development period.

Dr. William Johnson, of Search Technology, Inc. served as subject matter expert and schools liason for the SH-3 Bladefold system. He was responsible for the preparation of the system-specific information provided to the IMTS.

While Dr. Henry Halff had left the Office of Naval Research by the time the IMTS work was begun, his contributions to bringing the project into being were crucial.

NOTES

1. While impressive simulation systems were being developed and shown during this era, they were achieved by constructing highly specialized programs which embodied the behavioral rules of the target equipments. As a result these simulations could deal with time-variant effects and with immensely complicated equipment behaviors, but their range of application was limited to one system.
2. ESAS is a trademark of the Lockheed, California Company.
3. A conventional, frame-oriented instructional facility was added to GMTS/EEMT to provide some instructional content; however, this did not interact with the student during diagnostic practice, nor did it provide individualized assessment of diagnostic performance. This "programmed instruction" facility offered a reasonable way to implement an equipment overview which would be generally useful to all students.

REFERENCES

Bond, N.A., Jr., and Rigney, J.W. 1966. Bayesian aspects of troubleshooting behavior. *Human Factors 8:*377–383.

Bond, N.A., Jr. and Towne, D.M. 1979. *Troubleshooting Complex Equipment in the Military Services: Research and Prospects.* Los Angeles: Behavioral Technology Laboratories, University of Southern California, Report No. TR-92.

Edwards, W. 1968. Conservatism in human information processing. In B. Klienmuntz (Ed.), *Formal Representation of Human Judgment.* New York: Wiley.

Edwards, W. 1962. Dynamic decision theory and probabilistic information processing. *Human Factors 4:*59–73.

Hollan, J.D. 1983. STEAMER: an overview with implications for AI applications in other domains. Presented at the Joint Services Workshop on Artificial Intelligence in Maintenance, Institute of Cognitive Science, Boulder, CO, October 4–6.

Hollan, J.D., Hutchins, E.L., and Weitzman, L. 1984. STEAMER: an interactive inspectable simulation-based training system. *The AI Magazine* p. 2.

Johnson, R.A. 1960. An information theory approach to diagnosis. *Proceedings, Sixth National Symposium of Reliability and Quality Control in Electronics* January, pp. 102–107.

Kay, A. 1975. *Personal Dynamic Media.* Palo Alto, CA: Xerox Corp.

Manz, W. 1970. Experiments on probabilistic information processing. *Acta Psychologica 34:*184–200

Munro, A., Cody, J.A., and Towne, D.M. 1982. *Instruction Mode and Instruction Intrusiveness in Dynamic Skill Training*. Los Angeles: Behavioral Technology Laboratories, University of Southern California, Report No. TR-99.

Munro, A., Fehling, M.R., and Towne, D.M. 1985. Instruction intrusiveness in dynamic simulation training. *Journal of Computer-Based Instruction 12*(2):50–53.

Rigney, J.W., and Cremer, R.H. 1966. *Measurement and Prediction of Cognitive Loadings in Corrective Maintenance Tasks: I. A Bayesian Approach*. Los Angeles: Electronics Personnel Research Group, University of Southern California, Report No. TR-46.

Rigney, J.W., and Towne, D.M. 1974. Computer-aided performance training for diagnostic and procedural tasks. *Journal of Educational Technology Systems* 2(4):279–304.

Rigney, J.W., Cremer, R.W., Towne, D.M., and Bond, N.A., Jr., and Mason, A.K. 1967. *Measurement and Prediction of Cognitive Loadings in Corrective Maintenance Tasks: II. Bayesian Analysis of a Sample of Class B Electronics Technicians' Behaviors*. Los Angeles: Electronics Personnel Research Group, University of Southern California, Report No. TR-52.

Rigney, J.W., Towne, D.M., and Bond, N.A., Jr. 1969. *TASKTEACH: A Method for Computer-Assisted Learning of Serial-Action Tasks*. Los Angeles: Electronics Personnel Research Group, University of Southern California, Report No. TR-62.

Rigney, J.W., Towne, D.M., King, C.A., and Langston, E.L. 1972. *Computer-Aided Performance Training for Diagnostic and Procedural Tasks*. Los Angeles: Behavioral Technology Laboratories, University of Southern California, Report No. TR-70.

Rigney, J.W., Towne, D.M., and King, C.A. 1974. *Interactive Computer Graphics for Performance-oriented CAI*. Los Angeles: Behavioral Technology Laboratories, University of Southern California, Report No. TR-73.

Rigney, J.W., Towne, D.M., King, C.A., and Moran, P.J. 1978. *Field Evaluation of the Generalized Maintenance Trainer-Simulator: I. Fleet Communications System*. Los Angeles: Behavioral Technology Laboratories, University of Southern California, Report No. TR-89.

Rigney, J.W., Towne, D.M., Moran, P.J., and Mishler, R.A. 1978. *Field Evaluation of the Generalized Maintenance Trainer-Simulator: II. AN/SPA-66 Radar Repeater*. Los Angeles, CA: Behavioral Technology Laboratories, University of Southern California, Report No. TR-90.

Slagle, J. 1963. A heuristic program that solves symbolic integration problems in freshman calculus. In E.A. Feigenbaum and J. Feldman (Eds.), *Computers and Thought*. New York: McGraw-Hill.

Towne, D.M. 1969. Recursive functions for analysis of symbolic representations of task structure and human performance. Unpublished doctoral dissertation, University of Southern California.

Towne, D.M., 1979. The automated integration of training and aiding information for the operator/technician. *Proceedings, Third Biennial Conference on Maintenance Training and Aiding, Orlando, FL.*

Towne, D.M. 1982. Diagnostic psychological issues and maintenance design features. *Proceedings, Human Factors Design for Maintainers Conference, Pensacola Beach, FL.*

Towne, D.M. 1984. A generalized model of fault-isolation performance. *Proceedings, Artificial Intelligence in Maintenance: Joint Services Workshop, Boulder, CO.*

Towne, D.M. 1985. *Cognitive Workload in Fault Diagnosis*. Los Angeles: Behavioral Technology Laboratories, University of Southern California, Report No. TR-107.

Towne, D.M., and Rigney, J.W. 1979. *A Developmental Microprocessor-based System for OJT and JPA Management in Electronics Maintenance*. Los Angeles: Behavioral Technology Laboratories, University of Southern California, Report No. NAVTRAEQUIPCEN 76-C-0023.

Towne, D.M., Johnson, M.C. and Corwin, W.H. 1982. *A Technique for Projecting Maintenance Performance from Design Characteristics*. Los Angeles: Behavioral Technology Laboratories, University of Southern California, Report No. TR-100.

Towne, D.M., Munro, A., and Johnson, M.C. 1983. *Generalized Maintenance Trainer Simulator: Test and Evaluation in the Laboratory Environment*. Los Angeles: Behavioral Technology Laboratories, University of Southern California, Report No. NPRDC TR-83-28.

Towne, D.M., Johnson, M.C. and Corwin, W.H. 1983. *A Performance-Based Technique for Assessing Equipment Maintainability*. Los Angeles: Behavioral Technology Laboratories, University of Southern California, Report No. TR-102.

Weis, R.L. 1963. Bayesian inference and electronic troubleshooting. *Proceedings of the Operations Research Society of America, Honolulu.*

REACTIVE LEARNING ENVIRONMENTS FOR TEACHING ELECTRONIC TROUBLESHOOTING

John Seely Brown and Richard R. Burton

ABSTRACT

This paper describes the educational philosophy that we have evolved working on training and educational systems over the last 15 years. The description is largely in terms of the motivation for and high-level descriptions of many of the ideas evolved in and from the Sophie project—a project to explore the use of new computer and artificial intelligence technologies to teach electronic troubleshooting. Although these systems are fairly old by computer standards, the ideas they exemplify have renewed timeliness because advances in technology have produced inexpensive, powerful machines capable of providing the computational basis for their widespread application. Previous papers about Sophie stressed the artificial intelligence and computer science techniques that comprised its "intelligence." Many of the pedagogical assumptions, ideas, and methods that guided us during the develop-

Advances in Man-Machine Systems Research, vol. 3, pages 65–98.

ISBN: 0-89232-659-X

ment of Sophie I, II, and III were never addressed. However, during the last decade there have been major advances within the cognitive sciences that provide a useful framework for discussing the learning activities supported by Sophie. This framework allows us not only to better articulate the underlying intuitions but also to provide a more coherent rationale for these learning activities. To present these intuitions and rationales, we will discuss the kind of training we were aiming for and describe some of the cognitive, pedgogical, and philosophical ideas in Sophie that we hope will be helpful to designers of future training systems.

I. INTRODUCTION: LEARNING WITH UNDERSTANDING

A. What Do We Mean by "Learning with Understanding" and Why Is It Important?

Machines in the late twentieth century are becoming increasingly prevalent and increasingly complex, often involving a matrix of different interacting technologies such as electronics, mechanics, magnetics, and optics. Each of these in itself is difficult to understand; their interactions only increase the difficulty. As these systems grow in variety and complexity, the task of training individuals to install, maintain, calibrate and troubleshoot them becomes decidedly difficult and in some cases becomes the limiting factor on what can be realistically deployed—a problem all too familiar in the military and one that is beginning to get recognition in industry.

This paper is about training people to *think* in the context of maintaining complex pieces of equipment. Our goal is to consider what sort of training is required to produce troubleshooters who understand the equipment they work on. At present, many troubleshooters are trained to repair only specific problems on specific pieces of equipment. These people have great difficulty fixing unusual faults in familiar equipment or dealing with equipment for which they have no specific training. By contrast, a troubleshooter who can reason about how devices function can often find the one-of-a-kind fault and can fix even unfamiliar pieces of equipment. Such expert troubleshooters are often called in when others have failed. They must have sufficient conceptual understanding to develop diagnostic steps on their own and to digest information from technical manuals. At times they must be able to reason from first principles, transcending experiential knowledge gained by themselves or by others in the community of working technicians. It is around this kind of troubleshooter that we have oriented our training philosophy.

One key to robust troubleshooting is the ability to improve—to figure out the unusual, to recognize conflicting data and decide what should be discarded, and to know what to do when the appropriate instrument is not around. This knowledge stems from the interweaving of three kinds of knowledge: *what* knowledge, *how* knowledge, and, crucially, *why* knowledge. For example, in learning procedural skills, such as calibrating a test instrument, one must certainly understand both what is required and how to perform each step. Ideally, however, one should also understand the rationale behind each step in the procedure, that is, why the step is crucial or, if it is not crucial, why not. Such *why* knowledge helps troubleshooters to improvise when they get stuck following an established procedure. While the ability to improvise is widely useful in technical maintenance and repair, it is especially crucial in handling casualty situations since (1) much of the test equipment is apt to have suffered the same fate as the system being repaired and (2) there can be multiple and massive faults that radically alter how the system is currently functioning, rendering past experience with it either useless or, worse, deeply misleading.

B. What Kinds of Procedures to Teach

Troubleshooting procedures are standardly taught either as objects to be rotely learned or as objects that magically emerge from abstract engineering theory. Both of these (caricatured) positions have shortcomings.

Completely specified procedures might seem easy and cost-effective to learn—certainly easier than theory—but they are inherently brittle. They cannot be readily adapted to new equipment (sometimes not even to updates of the same equipment), to unanticipated operating situations, or to novel casualties. Furthermore, for a rote procedure to work in the field, it must be recalled with 100 percent accuracy. This is contrary to the nature of human memory. Human remembering is a problem-solving process that intelligently glues together fragments of partially remembered knowledge to "reconstruct" a coherent whole (Williams and Hollan, 1981). Since a rotely learned procedure has no underlying logical structure, the reconstruction of forgotten parts of it is blocked.

The standard alternative to the shortcomings of rotely learned procedures has been to teach "theory." But what kind of theory? Usually technicians are taught a watered-down version of the abstract theory taught to engineers. But troubleshooting seldom, if ever, requires that kind of analytic knowledge.

In contrast to these two approaches, we advocate teaching technicians (1) a kind of conceptual knowledge that is based on qualitative, causal models of how a system functions; (2) how to derive such causal models from the structure of the device; and (3) how to use that knowledge to understand the rationale behind various troubleshooting procedures. The qualitative, causal knowledge and related reasoning strategies underlying this approach enable the technician to construct mental models of how and why the device functions as it does. Metaphorically, these models can be viewed as qualitative simulations that *run* in the *mind's eye*.

C. What's So Hard About Electronics?

What is it about electronic circuits that makes them hard to understand and troubleshoot? Feedback. Nearly all useful circuits contain feedback, and systems that use it are substantially harder to troubleshoot than those that don't, such as the typical battery-lightbulb-and-switch circuits used in elementary electricity courses.

The Sophie laboratory, which we describe later, works with moderately complex devices that contain multiple and complex feedback loops. On such systems, naive students try to apply strategies such as the classical half-split method to linear chains of components within the circuit. This strategy works well when troubleshooting simple circuits that use no feedback to achieve their overall function. But in devices that use feedback mechanisms to control their own behavior, the application of the half-split technique *to components*—i.e., starting in one place and moving in a linear fashion through the circuit—fails, because a bad measurement anywhere in the feedback loop might be caused by a fault in *any* of the loop's components. This failure arises because of the dual functions enacted by feedback circuit and the difficulty of distinguishing the behavior caused by them. Specifically, feedback circuits can often be thought of as containing both a producer that generates voltage or current and a compensator that regulates the generated quantity to the desired level. If one is to interpret the fact that a measurement between the two is low, one must first decide whether it is a result of the producer not generating enough or of the compensator overcompensating. In feedback cases, then, a model of how the overall device works is crucial because it is to the *causal chains* that the half-split method must be applied, and these do not necessarily stand in a one-to-one relationship to the components. Said differently, one must try to halve the space of possible faults, not the space of components.

Note that feedback is more than just a loop in the circuit's schematic. All circuits have loops; without a loop, no electricity flows and nothing at all happens. Feedback is a characteristic of the *causality* of the circuit and may or may not involve a structural loop. Feedback usually requires active elements such as transistors, op-amps, etc. Often feedback can not even be detected without first tracing through all the causal pathways of a circuit and then coalescing them into a coherent overall model of how it functions.

D. Mental Models: More Than Simulation

The mental model of the device that an expert troubleshooter needs is more than just a qualitative, causal simulation of it. In addition, he must understand how the particular mental model relates to (or is derived from) the actual structure of the device, from its actual components and their interconnections; he must be able to move from the structure of the device to its function. Without this ability, he can't predict how the observed symptoms might be explained by perturbations or faults to the various components of the device. Thus the expert must augment his mental model of the device's causality either by explicitly encoding how fragments of the model relate to fragments of the structure or by being able to derive this relationship on the fly, using a problem-solving process called envisioning. Interestingly, by being able to construct the relationships between structure and function through the envisioning process, the expert can even handle faults that fundamentally alter how the device is functioning, faults that normally would render one's mental model of the working device useless. (See de Kleer and Brown, 1983, for further discussion on this point.)

E. Troubleshooting as a Knowledge Integration Activity

Much of what follows focuses on teaching electronic troubleshooting. To us, teaching troubleshooting is as much a pedagogical activity for crystallizing a student's knowledge of electronic theory as it is a goal in itself. Theory alone does not constitute a useful knowledge of electronics; it must be combined with practical experience. Indeed, it is the purpose of theory to unify and organize the vast experience of many people in a form that is easier to communicate and to use in support of making predictions. But a student who has little experience before learning electronic theory will learn facts and relationships as an abstract body of knowledge. When such a student begins working in field situations, he has a great deal of conceptual relating and reorganizing to do. One of the goals was to have Sophie provide just the sort of experience that makes theory meaningful. Even if knowing how to troubleshoot were not a primary goal for electronics students, learning how to troubleshoot is invaluable because of the depth it imparts to their theoretical knowledge.

II. REACTIVE LEARNING ENVIRONMENTS

A reactive learning environment is one that responds to a user's action or request in a manner that extends his understanding of his own action in the context of the specific situation. Reactive learning environments are not just interactive, graphic, or analytic simulations. They often use simulations, augmenting them with an intelligent coaching system that can help focus a student's attention on the relevant aspects of how the system is responding to his actions. To see why more than simulation is often needed in a learning context, we need first to consider the role of the student's perceptual mechanism in seeing or ''parsing'' the behavior of the simulation and then how that mechanism can be affected by a student's conceptual system.

A. Simulations and Story Construction

A key use of simulation of a complex device is to enable the student to observe how the system functions and to witness how its behavior changes in response to changes in the state of one of its components or changes to its external variables or inputs. But the behavior of the simulation must, in its own right, be interpreted by the student. Take a simple case in which a simulation animates the spreading effect of a change to the system. That animation reflects a causal chain of events but is not, in itself, a causal model. The causal (mental) model must still be constructed by the student through interpretation of the observed behavior of the simulation. Thus, a given event must be interpreted as *enabling, inhibiting,* or *triggering* another event, which in turn enables, inhibits, or triggers still other events, and so on. Furthermore, there must be some medium through which information is "communicated" that enables events to trigger or inhibit other events. For example, a simple-minded animation of electrons flowing around a circuit can be interpreted by the student as supporting either a fluid flow model or a crowd flow model of how information is communicated around the system. As Gentner and Gentner, 1983, point out, these alternate models can lead to different predictions about behavior.

In general, then, students must construct some kind of causal story for the simulation's manifested interactions of the various components of the system. But a story is constructed out of things perceived, and perception is expectation-driven. That is, students will only see epistemological distinctions and events that they already know about or that are at least consistent with their current view of how the world works. Speaking metaphorically, unless the student's conceptual eyeglasses are more or less attuned to the world view or ontology of the simulation designer, then what appears as a meaningful event to the designer may be seen either as noise or not seen at all by the student. This possibility is not just academic. Experiments with the Logo-based Dynaturtle (di Sessa, 1982; White, 1984), which provides a vivid simulation of a frictionless, Newtonian world (see Figure 1), reveal that, when confronted with dramatic counterexamples to what they believe will happen when they "kick" a simulated frictionless puck, kids simply do not "perceive" the contradiction to their cherished beliefs.

The experience with the Dynaturtle is not an isolated event. In the last few years there have been many studies demonstrating the inability of formal training to change students' conceptual, commonsense view of how the physical world functions (Clement, 1983; McCloskey, 1983). Students can "understand" Newton's laws in the sense that they can use them to solve the neat textbook problems at the end of each chapter in their physics textbook. But even after students have successfully finished an entire year of classical mechanics, they often fail to build a connection between this formal knowledge and their commonsense knowledge or belief about how the world around them actually behaves. The conceptual eyeglasses that they use to interpret physical events have not changed. They look at events happening in the real world and see real objects, not the idealized objects of mass, acceleration, momentum, energy, etc. of Newtonian physics. Their formal knowledge remains detached from their experiential knowledge.

For meaningful learning to take place, this chasm must be bridged. Somehow, hypotheses derived from the entailments of formal knowledge must be connected to and evaluated against students' experiential knowledge. In this way, their intuitions about the physical world can slowly change and, in turn, permit a deep reconceptualization of their theoretical understanding of the physical world and its underlying mechanisms.[1]

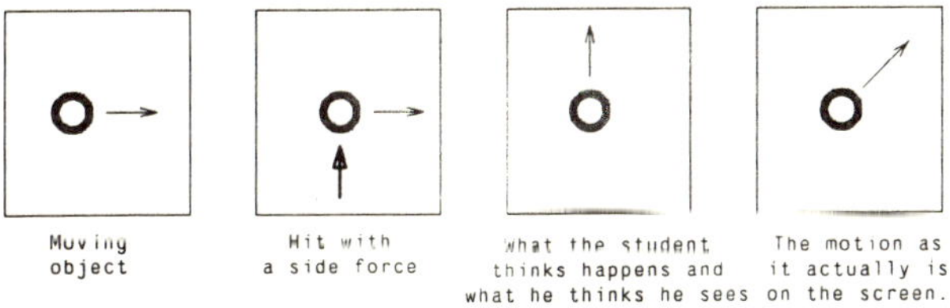

Figure 1. Example of a student's belief blocking what he or she sees.

Stated most ambitiously, the goal of reactive learning environments is to help bring about a reconceptualization of, in our case, the world of electronics. This reconceptualization is encouraged by getting students to actively form hypotheses about the entailments of the behavior they observe in the simulation and to have those hypotheses critiqued by an ''intelligent'' hypothesis evaluator. Theory is thus brought in direct and often dramatic contact with experience in a way that makes it difficult not to see.

B. Learning as Reconceptualization

A common image of education is that of a teacher pouring stuff (knowledge) into an empty vessel (the student's head). This image arises from viewing learning as an *assimilation* process. Models of learning currently being developed in the cognitive sciences view learning as a *reconceptualization* process. From this perspective, learning is more like a series of miniparadigm shifts. For example, the novice in electronics starts out with a conceptual system, one he most likely invented by using the available explanatory mechanisms stemming from his understanding of his physical and social worlds. From there he gradually constructs a new conceptual system that does justice to the formal knowledge that he is taught and that, hopefully, also connects to his experiential knowledge.

A conceptual system is more than just a collection of laws and technical definitions. Like Wiser and Carey (1983), we think of a conceptual system as involving three major dimensions: (1) the phenomena being explained or the problems and tasks being undertaken; (2) the laws and definitions comprising a theory of the domain; and (3) the models, the conceptual entities that are the models' constituents, and the set of mechanisms that enable the model to communicate information between its constituents, i.e., enable the model to ''run.''[2]

From our perspective, there is a crucial difference between the a domain's theoretical laws and its models of phenomena. For example, knowing that Ohm's law is $E = IR$ does not mean that one knows the underlying models of current flow and potential difference that are united by this formal law. Indeed, in learning Ohm's law, one might avoid constructing any conceptual model of the domain at all, thereby sidestepping a semantics for this law and expecting that its applicability conditions can be learned in a purely syntactic way. Although such an approach might work for very simple problems, it is easy to construct patho-

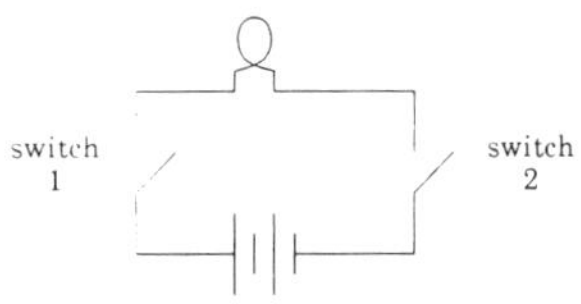

Figure 2. A simple electrical circuit. When both switches are open, the potential difference across either is not defined.

logical situations that violate most straightforward applicability conditions for theories even as simple as Ohm's law and that thus require a conceptual model to determine how to effectively apply them. For example, consider the example in Figure 2 involving two switches in series which are, in turn, connected to a light bulb and battery. When one of the switches is closed, then one can meaningfully measure the potential difference across the other one. When both are open, then the potential difference across either is ill defined. Depending on the model one uses to connect the concept of effort with the concept of flow, this fact is either obvious or nonsensical. For example, the standard fluid analogies for electricity in which pressure is analogous to voltage would lead one to think that the potential difference is well defined.

C. Engage and Pull: The Key to Reactive Learning Environments

As we explained, a student comes to a learning situation with a conceptual system or set of eyeglasses with which he makes sense of phenomena. In order to modify that conceptual system we must both engage it by setting it to work on real problems and pull it by demonstrating concrete weaknesses in the system. In Sophie I, hypothesis evaluation engages students by inducing them to articulate their hypotheses about what might be wrong with the device. The pull comes from the construction of explicit counterexamples to their hypotheses, counterexamples that are articulated in a different medium from the simulation.

Hypothesis evaluation is one of the primary pedagogical techniques of the Sophie I system. To understand the hypothesis evaluation system and its overall pedagogical role, it is helpful first to have an idea of how Sophie I is used. In a typical session, a student is presented with a regulated power supply that contains a fault of some specified degree of difficulty. He then tries to debug it by requesting measurements from Sophie's simulation/workbench. At any time he can offer a hypothesis about what he thinks might be wrong with the device and receive an evaluation of his hypothesis relative to the measurements he has thus far made. That is, the system tells him not whether his hypothesis is right or wrong but whether or not it is *logically consistent* with the information he should have learned about the device from his measurements. At any time, the student may also request that a component be replaced. Before the part is actually replaced, the student is required to say exactly what he thinks is wrong with the part. Only if he is correct is a new component put into the circuit. If he is incorrect, the hypothesis evaluator is invoked to determine whether his proposed fault is reasonable and if not, why not.[3]

III. SOME ISSUES IN PEDAGOGY

In our early experiences with Sophie I, we recognized a need to introduce students to how to use facilities provided by the system. The students knew a lot about electronics but were not familiar with the particular device being studied. Additionally, although the students had done troubleshooting before, few had been taught to troubleshoot. As part of Sophie II, we designed an articulate expert troubleshooter that was to serve two functions. The first function was to demonstrate to the student how to use the full range of functional capabilities of the Sophie workbench by actually using them to locate a fault chosen by the student. The second, more important, function was to provide strategic troubleshooting knowledge to the student in a context from which he might maximally profit. In brief, the system first models an expert troubleshooter, articulating the thinking processes that lead to all of the decisions. It then pulls the student into playing part of its troubleshooting role by turning over certain tasks to the student, and finally it fades away completely, becoming just a limited coach watching over the student as he tries to perform similar tasks.

A. The Articulate Expert

The student's first interaction with the system proceeds as follows: An agent referred to as the ''referee'' introduces itself and interacts with the student, letting him choose how the circuit should be faulted. The referee then inserts the student's fault in the circuit and calls upon the ''expert'' to locate it. The expert decides what measurements to make, just as the student will be expected to do later when he is troubleshooting. The reasons for taking each measurement that is requested by the expert are explained to the student in terms of various debugging strategies based on a qualitative analysis of the circuit. Thus, the student gets a chance to hear why the expert is making a particular measurement and what might follow from knowing its value. Before the expert actually makes each measurement using the Sophie lab, the student is asked to predict the *qualitative* value of the measurement (will the measured value be too high, too low, or about right?). This forces the students to propagate the symptoms of their chosen fault through other parts of the circuit. The debugging process continues with the expert making measurements until it has isolated the fault.

We want to stress that the importance of the articulate expert is *not* that it can locate faults in the given device, but rather that it can articulate why it chose to make a particular measurement and what one could deduce from knowing its value. That is, it can explain its particular troubleshooting strategy, why it is making any particular measurement, what follows from the results of that measurement, and what someone might mistakenly or prematurely conclude from that measurement.

Unlike most artificial intelligence (AI) systems, the articulate expert foregoes the goal of generality in order to attain cogent explanations, especially those pertaining to strategic issues. The underlying mechanisms of the expert are based on parameterized decision trees annotated with schemata for producing explanations about troubleshooting strategies. From an AI technology point of view, this is very limited. But it does have pedagogical advantages. Since the annotations are prestored and crafted by hand, they can be made as insightful as desired. Furthermore, since the annotations are associated with nodes in an augmented decision tree and since any node can be reached by only one path, their exact context is known ahead of time. Thus, the explanation can be specific to the situation at hand, stating misconceptions and examples to overcome them that are relevant just here. Each node has several schema, with the choice depending primarily on how often the student has activated this node.

The expert is not committed to a single troubleshooting strategy, but will follow one of several depending upon the situation. By offering multiple approaches, the student is exposed to alternative ways of attacking the problem—letting him witness an expert that is flexible enough to use different strategies for specific reasons. For example, in a situation in which the output voltage of a reulated power supply is high, the two following strategies are candidates:

1. *Milk the front panel*—Extract as much information as possible from observations of the external behavior of the device under different settings and loads before proceeding to internal measurements. This strategy has an important caveat: Don't make things worse. To keep from blowing the circuit when ''milking'' the front panel, be sure not to set the controls and put a load on the system that would ordinarily invoke the protective mechanisms of the circuit, because these may not now be functioning.
2. *Hypothesis splitting*—Formulate some, though not necessarily all, possible hypotheses and take measurements that would distinguish between them. In particular, don't spend a lot of time convincing yourself that you have considered all possible hypotheses before checking out the more likely of the ones you have generated.

There are several pedagogical reasons for conducting a dialogue between the student and the expert. The first is to convey useful debugging strategies to the student and to demonstrate the usefulness of causal models of the circuit. The expert views the device in terms of discrete boxes which interact in known ways. It has, in other words, a high-level causal and teleological model of how the circuit works.[4] Without having to make measurements inside functional blocks themselves, the expert is able to determine which block contains the fault. Watching this process, the student sees the value in reasoning at this level.

A second reason for providing the expert debugging interaction is that it gives students the practice of envisioning the consequences of symptoms and faults. When the referee asks students to predict the value of a measurement that the expert is about to take, they are forced to use their causal understanding of electronics in general and of the circuit in particular. This process will often force students to consider the effects of their fault on parts of the circuit that they had not considered before. This skill is needed by the students when they are troubleshooting in order to check the predictions of their own hypotheses about what is wrong.

The expert also has a built-in strategy for turning over the task of predicting the consequences of a measurement to the student after initially letting the student observe.

B. The Articulate Expert as Reciprocal Teacher

We have only recently come to understand the generality and pedagogical advantages of the way we used the articulate expert to convey serious strategic, tactical, and conceptual knowledge of electronics and the attendant activity of troubleshooting. Palincsar and Brown, working in the domain of reading, have made the observation that particular kinds of involvement with an articulate expert is a powerful general teaching technique (Palincsar and Brown, 1985; Brown and Palincsar, 1986).[5] They designed a teaching method employing a human expert that has proved remarkably effective in raising students' scores on reading comprehension tests, especially those of poor readers. Their basic method centers on demonstrating and having students employ the strategic skills used by an expert reader. The method is called *reciprocal teaching* because the teacher and students take turns playing the role of teacher.

The teaching procedure is as follows: Both the teacher and student(s) read a paragraph silently to themselves. Whoever is playing the role of teacher formulates a question about the paragraph, constructs a summary, and makes a prediction or clarification about the text. Initially, the teacher performs this process in order to give the students a model of what an expert reader does. Gradually the tasks are turned over to the students (who take turns if there are more than one). When students first undertake the process, the teacher coaches them extensively on how to construct good questions and summaries, offering prompts and critiquing their efforts. In this way, the teacher provides a "scaffolding" for the students, enabling them to take on whatever portion of the task they can. As the students become more proficient, the teacher turns over more of the task to them. At any point, student and teacher can switch roles so that the student becomes the modeler or critiquer with respect either to the teacher or to other students in the group. This reciprocity, expressed either through role switching or through subtle adjustments in the scaffolding, is a key ingredient of the method.

There is a great deal of experimental data that support the claim that this method is extremely effective, but why is it so effective? Its effectiveness appears to depend upon the co-occurrence of a number of factors.

First, the method teaches students a wholly new idea about the nature of reading as a skill—i.e., that reading requires actively formulating questions and making summaries, clarifications, and predictions, rather than simply scanning text or saying it aloud.

Second, the teacher *models* expert strategies in a problem context shared directly with the student, which encourages students to reflect on the teacher's performance. Thus, the teacher articulates for the student the questions he would ask about a paragraph that they have both just read. The students listen in the context of knowing that they will soon undertake the same task, so they attend closely to what he is doing. After they have tried to do it themselves and perhaps had difficulties, they listen to the teacher with new knowledge about the task. As they read a passage silently, they may try to generate a question or summary themselves, noticing later what he does differently. That is, they can compare their own question or summaries with the questions or summaries he generates. They can then reflect on any differences, trying to understand what led to those differences.

Third, the method of scaffolding gets students started in the new skills, giving them confidence that they can do them (Meichenbaum, 1985). For example, in formulating questions, the teacher might first want to see if the student can generate a question on his own; if not, he might suggest starting a question with "Why" or "How." If the student still can not generate a question, he might suggest formulating a simple "Why" question about the agent in the story. If that fails, he might generate one himself and ask the student to reformulate it in his own words. Scaffolding is designed to help students when they are at an impasse. With successful scaffolding techniques, the student gets as much support as he needs to carry out the task, but no more. This slowly builds confidence in the student that he can master the required skills.

The final important aspect of reciprocal teaching is fading, which occurs as the teacher turns control over to the students. Fading forces students to take over more and more of the task, so that they develop all of the skills needed for the task. In particular, they are forced into the dual roles of *producer* and *critic*. By becoming critics as well as producers, students are forced to articulate their knowledge. They cannot just say that a particular summary is bad. They must say what makes it good or bad. Articulating knowledge is a step toward freeing it from its contextual binding, so that it can be used in many different contexts.

There are many parallels between reciprocal teaching in reading and the role played by Sophie's articulate expert in electronic troubleshooting. Like the reciprocal teacher, the expert provides a model of how to think about circuits and how to use different types of knowledge in troubleshooting, which is new to most students. It demonstrates the importance of certain activities—formulating hypotheses and following their consequences. And the expert questions the students, engaging them into the troubleshooting process.

Like reciprocal teaching, the design of the articulate expert does fading when, after several interactions, the expert asks the student to make some of the conclusions that it normally would make. This gradually pulls the student into carrying out the expert's tasks. The ultimate fade is when the student is presented with the Sophie Lab experience on his own.

An effect similar to scaffolding is achieved by alternating between the expert and the laboratory activities. When the student first sees the expert's explanations, he may quickly read through them without fully comprehending what is being said. Following this, the student is given a fault that he must find. In those cases where the student has an incomplete or inaccurate model of the circuit, he will be confronted with a set of seemingly contradictory observations. After the student has struggled his way through, he can give the same fault that caused him difficulty to the expert and watch what it does. When the expert runs across the same contradictory observations, the student pays heightened attention to its method of successful resolution. Often the escapes from apparent dilemmas focus the student's attention on details or complexities previously overlooked. Conversely, some things that concern the expert—reasons for inferences, caveats, and the like—may only make sense to the student later during a debugging problem that he can solve by appealing to them. Thus, the combination of the expert and lab exercises is more effective than either would be in isolation.

C. Games That Improve Debugging Skills

There are also aspects of reciprocal teaching in another Sophie II activity: the two-person debugging game. The two-person debugging game was developed to take advantage of the competitive nature of some students to motivate more thoughtful troubleshooting. In the game, one player (the *inserter*) chooses a fault that another player (the *debugger*) must then find. The game begins with the inserter introducing the fault of his choice into the circuit and setting the front-panel controls so as to exhibit an external symptom. That is, the inserter must set up the device so that it is malfunctioning. The debugger then finds the fault by performing a sequence of measurements. Each measurement has an associated cost that roughly reflects the degree of difficulty of making that measurement on a real instrument. For each measurement the debugger makes, the inserter must predict whether the result of the measurement will be higher, lower, or approximately the same as it would be in a working instrument. After the debugger successfully locates the fault, the inserter is given a score of the total cost of the debugger's measurements multiplied by the percentage of times the inserter correctly predicted the outcome. Thus, it is to the advantage of the inserter to choose a difficult fault but not one so difficult that its consequences can't be predicted. The debugger, of course, attempts to isolate the fault using the least expensive sequence of measurements. The debugger announces a guess by replacing the component believed to be faulted. If the wrong component is replaced, the cost is increased by a substantial amount. After each successful fault discovery, the roles are reversed and another game initiated.

One noteworthy aspect of the game is that it forces the inserter to use his knowledge of electronics to reason forward from causes to symptoms. In teaching debugging, the primary emphasis is typically on reasoning backward from symptoms to causes. However, even in debugging, forward reasoning is important to evaluate hypotheses and make sure that they do explain all of the observed symptoms and that they would not generate any symptoms that have not been noticed. Playing the role of inserter requires forward reasoning in two ways. First, when the inserter chooses a fault, he must reason forward from that fault to determine a set of instrument controls that exhibit a symptom. Second, whenever the debugger makes a measurement, the inserter must predict the behavior of the system at the measured point. This requirement has the possibility of substantially deepening the inserter's knowledge; when the measurement is off the direct causal path from the fault to the symptom, he will be forced to think about parts of the circuit he otherwise might not have bothered to master.

An additional benefit of the game is that it puts a premium on fast, accurate reasoning about the circuit. This encourages the student to exercise his causal and teleological understanding of the device. Since the causal and teleological models of the device capture the group behavior of collections of components, working at this level is much faster.

D. The Team Games as Protocol Gathering and Reciprocal Teaching

Extending the debugging game from a two-person game to a team game, with each team comprised of two people, leads to extraordinarily productive interactions. Teams are more adventuresome, more willing to probe the limits of their understanding. The discussion of possible ramifications of alternative candidate faults that arises while choosing a fault, for instance, often requires substantial pooling of the inserting team's members' knowledge as well as correction of each other's misconceptions. Similarly, when deciding on which next measurement to make, the debugging team will often discuss the possible ramifications of each suggested measurement.

Observation of the teams is a valuable way to gather protocols for cognitive research. Group protocols can provide a natural "window" into the strategic reasoning patterns and knowledge of the individuals on the team (Brown, Rubenstein, and Burton, 1976). The discussions or arguments that unfold between team members provide detailed evidence about their strategic reasoning, including not only the reasons that a given measurement is selected but also the reasons that some other is not selected. Observation of teams also helps solve another problem plaguing the use of protocols in studying reasoning—getting subjects to articulate what they are thinking without becoming self-conscious and thereby altering their natural problem-solving behavior. We find that team players are far less self-conscious than the single informant and that they make and defend hypotheses, plan and revise strategies, and explain things to each other in what appears to be a natural way. Likewise, in collecting a protocol of a subject who is working alone, it is extremely difficult to get insights into why he rejects certain moves; subjects usually feel no need to justify why they *don't* do something. Thus, for example, the researcher usually has no idea what, if any, other measurements the student considered. Instead, single subjects invariably focus on the more tactical issues of how to accomplish their objects. In the team environment, articulation of reasons and strategies naturally arises in the attempt to convince one's teammate of the validity of a proposed move. The record of these justifications provides a unique opportunity to see strategic reasoning unfold and be defended.

The same factors that make the interactions between team members useful for the researcher studying strategic reasoning also make them a good reciprocal teaching process for the team members themselves. Consider the task of the debugging team. Each member of the team must argue for selecting a particular measurement over his teammate's. Thus one must be able to select a good measurement, articulate why it is a good measurement, and critique someone else's reason for why their measurement is better than that of his teammates. Moving the process of choosing a measurement from inside a student's head to being argued out in a group forces the reasons for a decision to become articulated. When these reasons are articulated, the members of the team who don't have them can get them and the members that do have them can disagree (with an articulate counterargument). This is exactly the stuff of reciprocal teaching, with expertise distributed among team members.

A similar opportunity is presented to the inserting team. Here a fault must be selected. It must be difficult to troubleshoot, but its consequences must not be impossible to predict. This requirement means that the inserted fault must be thoroughly envisionable, at least for those measurements that the opposing team actually makes. Furthermore, if one can predict the troubleshooting strategies of the debugging team, it may be possible to choose a fault that is hard for those strategies to find. Thus, forward reasoning about circuits, opponents' troubleshooting strategies, and their limitations are all things that can be argued out explicitly in a manner similar to the reciprocal training method. (We note here that this form of reciprocal teaching will not work in all situations. In particular, if none of the students knows anything, there is no "expert" behavior for the others to mimic. Thus, this is a good activity for spreading knowledge between students who know some electronics.)

E. Explanatory Sequences

We mentioned earlier the importance of teaching the student to infer the function of a circuit from its structure. That is, the students should be able to pick up a schematic and figure out what would happen if any part were faulted. The failure mode of a part may not be as simple as opening or shorting. For example, there may be a shift in the beta of a transistor or a change in the value of a resistor. To be able to predict the consequences of these subtle faults, one must understand why each part is in that circuit. That is, the student must know what role that part actually plays in the overall functioning of a circuit and what mechanism it exploits to achieve its role in that overall plan. For example, a resistor whose purpose is to provide a voltage drop around a feedback circuit may change its resistance by 100 percent without affecting the circuit's behavior, while one whose purpose is to sense current flow may exhibit symptoms if it changes by 5 percent.

One very powerful way to teach this knowledge is based on the intuition that we can understand a complex circuit module by understanding how the basic function of that module can be most simply realized and then recognizing a sequence of evolutions or patches to that simple realization that take care of its shortcomings. Thus, the presentation takes the form of an explanatory sequence of circuit designs. The sequence begins with a drastic simplified circuit that captures the central part of the desired behavior. An explanation of what the simplification successfully captures is given and, more importantly, the reasons why it does not satisfy all of the requirements are detailed. A refinement to the design is then proposed to create a slightly more complicated hypothetical circuit. This process is repeated, resulting in an evolutionary sequence of circuits which finally converges on the real circuit. Each successive version has an identified shortcoming and a way to compensate for it by adding a new component or module.

We believe that such evolutionary sequences lead to a better intuitive understanding of the role of each component in the real instrument than does the usual sort of straightforward functional description found in technical manuals. In an evolutionary sequence, each functional block undergoes its own evolution from primitive to realistic. In this evolution, the purpose of the module in the actual circuit becomes readily apparent, as are the mechanisms by which its purpose is realized. That is, both its intrinsic and extrinsic descriptions become understandable.

To make the above discussion more concrete, we will outline the evolutionary sequence for one of the modules from Sophie. The example is shown in detail in Brown, Rubinstein, and Burton (1976). The sequence explains the constant-current source module of the power supply, which is the most difficult module for students to understand. The pieces of the circuit that make up the constant-current source are shown in Figure 3.[6] The explanation begins with a description

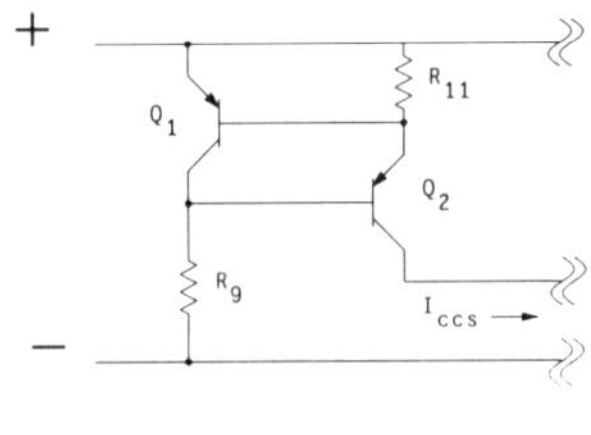

Figure 3. The constant-current source module.

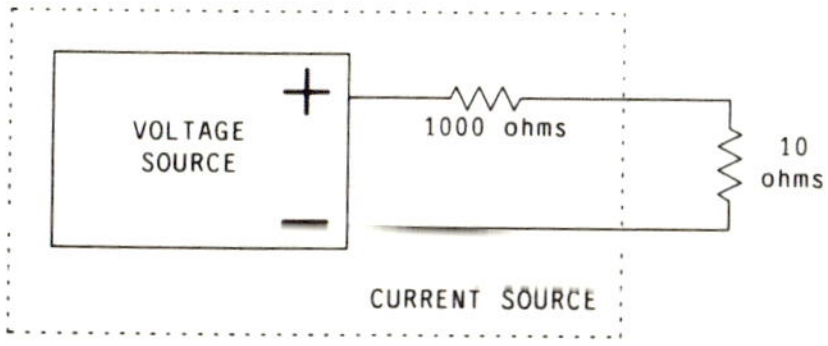

Figure 4. A simple current source.

of how a large resistor in series with the load acts to maintain a relatively constant current over a range of output loads (see Figure 4).

The shortcomings of this approach are then explained: The current is not really constant, its variability is just greatly dampened; and it uses a large amount of power.

The second circuit stage (shown in Figure 5) replaces the resistor with a transistor (Q_2) biased as a linear amplifier. The explanation describes the electrical properties of the transistor that produce this linear amplification. This circuit is better than the first one in that it reduces the power requirements and produces a more constant current. However, a change in output load still results in a small change in current because of the difference between the ideal model of how a transistor behaves and how it actually behaves. Also, this circuit has a problem because it responds slowly to changes in the load, producing dips in the current.

The third stage replaces one of the resistors in the biasing network (R_x) of stage 2 with a second transistor (Q_1) that adds negative feedback, thereby correcting the deficiencies of stage 2. This is the final circuit shown in Figure 3.

One can imagine providing simulations of the intermediate designs that the student could explore in order to gain a more concrete understanding of each module's improvements and shortcomings. In the Sophie system, this was not done. Instead, the evolutionary sequence was used to organize the workbook materials that accompanied the laboratory and to organize the explanatory text generated by the articulate expert. (White and Frederickson (1986) have done some extremely interesting research on a typography for sequences of qualitative

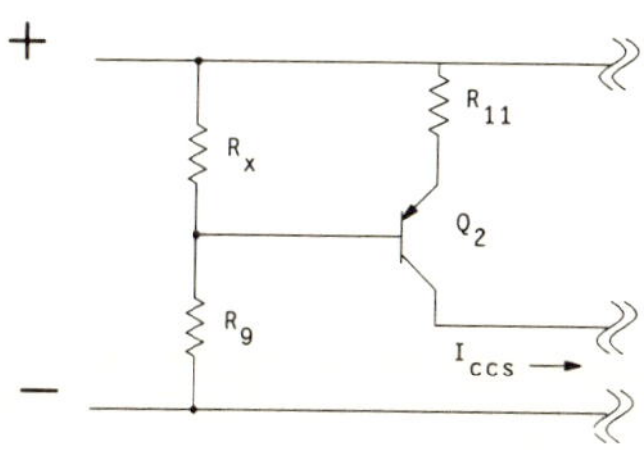

Figure 5. A better current source module.

models that provide successive depth of understanding of circuits. While Sophie aimed at a sequence of different circuits to explain a particular function, Quest provides a sequence of different understandings of the same circuit.)

IV. THE METHODOLOGY OF PARADIGMATIC SYSTEMS

Midway through the Sophie II project it became clear that in order to construct a good computer-based coaching system or automated laboratory instructor, we would have to make significant advances in the functional areas comprising coaching. Early on we had decided to decompose the functional components of a coach into four modules (see Figure 6): (1) *domain expertise,* which mirrors the way that experts reason in that field; (2) *diagnostic expertise,* which infers a student's strengths and weaknesses from his behavior; (3) *tutoring expertise,* which determines when to break into an interactive session and the content of what to say when interrupting; and (4) *explanatory expertise,* which communicates things in ways that match the student's particular world views, knowledge states, and cognitive styles.

Given that substantial progress was required on all four fronts, we devised a research strategy that allowed us to attack each module separately. That is, we built prototype systems in domains that minimized three of the components, enabling us to focus on just one in each system. This strategy led to a sequence of projects. The goal of the West project was to develop the tutoring or coaching expert module (Burton and Brown, 1982). The realm of an arithmetic drill-and-practice game was chosen because the domain expert was relatively simple; the diagnostic expert was straightforward, and the explanation expert did not need to worry about coupling into and/or shifting world views. More or less simultaneously, we began the Buggy project, in which the domain expertise (place value subtraction) seemed straightforward and coaching and explanation were not addressed at all. Instead, the task of interest was to develop a robust diag-

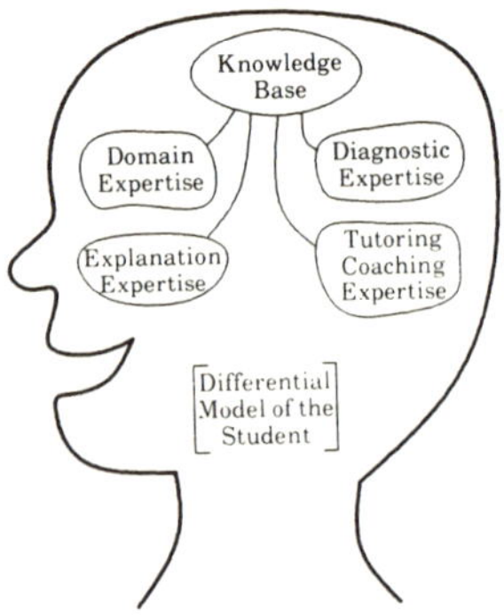

Figure 6. The structural components of a computer-based coach.

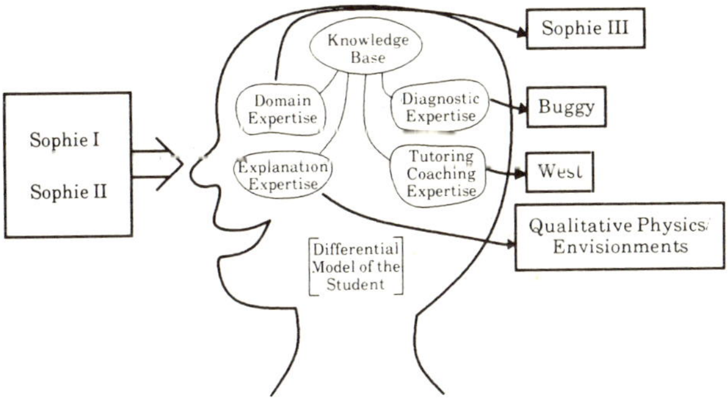

Figure 7. The structural components for a computer-based coach arose from work on Sophie I and Sophie II. The projects on the right each focused on a different component.

nostic theory and set of mechanisms that could construct from the student's behavior explicit procedural models of the student's misconceptions or bugs (Brown and Burton, 1978; Burton, 1982). The Sophie III project—which we discuss later—focused exclusively on constructing a domain expert for electronics and electronic troubleshooting. This expert system captures the set of intuitions we have about our own expertise at troubleshooting, particularly the way we reason about complex electronic circuits. Lastly, we began a project that focused on the notion of explanation, which continues to date. This work has led to the development of theories of qualitative physics and of the mental models produced by various kinds of envisioning processes. The relationships among these projects and the structural components of a computer-based coach are shown in Figure 7.

V. SOPHIE III: A HUMAN-LIKE TROUBLESHOOTING EXPERT

As we have just seen, one of the components needed to realize the goal of a computer-based coach is appropriate domain expertise. The goal of Sophie III was to develop an expert electronic reasoner and troubleshooter that more nearly mimicked the ways that human experts reasoned so that it could act both as the domain expert for a coaching system and as a foundation for inferring students' reasoning strategies and knowledge. One thing that makes this a difficult task is that reasoning about electronics and electronic troubleshooting is an exceedingly complex domain, only a small part of which has ever been formalized. We did not want to restrict ourselves to those parts that had already been formalized since those were the more mathematical aspects of the domain. We were primarily interested in qualitative and informal reasoning aspects of electronic

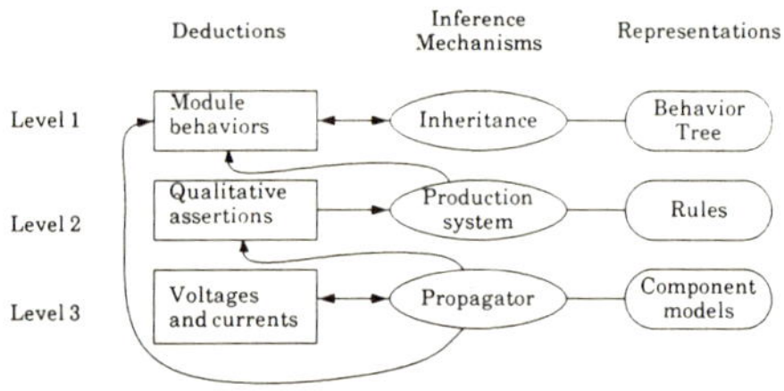

Figure 8. Diagram of Sophie III's electronics expert.

expertise that have barely been recognized, let alone formalized. On the other hand, we wanted this system to be as circuit-independent as possible, unlike Sophie I, and as robust as possible. These are contradictory goals.

de Kleer and Brown resolved this problem by breaking the reasoning components of the system into three levels, shown in Figure 8. The lowest level (level 3) captures generic electronic knowledge and reasoning strategies such as Ohm's law, Kirchoff's laws, and generic models of primitive components (e.g., transistors and diodes). At this level, the reasoner is completely circuit-independent and is based on a constraint propagation scheme that can propagate assertions or measurements throughout the circuit. This local propagator deals with ranges of quantities rather than single numeric quantities in order to represent the fact that any measurement is to some degree inaccurate, as are the values of all the components. The use of value ranges complicates constraint propagation since the system must propagate intervals and use interval arithmetic, but it makes it possible to handle many kinds of inferences that humans actually perform and provides a rudimentary notation for expressing *qualitative* constructs. The quantitative ranges produced by the local propagator are translated into qualitative assertions and used as input to the second level.

Levels 1 and 2 capture circuit-specific knowledge. This circuit-specific knowledge is organized around a functional decomposition of the circuit. A circuit is a designed artifact, consisting of a collection of weakly interacting modules. The modules behave cooperatively to produce the behavior of the overall circuit. The key to the circuit-specific knowledge is devising terms to describe the behavior of the modules such that the descriptions can be interfaced to the other functional modules.

Level 2 uses production rules to capture the circuit-specific knowledge that explains the behavior of the individual modules. It is aimed directly at those situations in which general knowledge fails. Because its context is limited and known, the rules can be put into a canonical form, the number of ad hoc rules can be minimized, and the determination of whether enough circuit-specific knowledge has been included is made much easier. The top level (level 1) captures the circuit-specific knowledge necessary to combine the behaviors of the modules into the overall behavior of the circuit.

The explanations of the circuit-specific knowledge are supplied by the electrical engineer and attached to each knowledge rule. In contrast, explanations for deductions made from the local propagator can be constructed from the generic electronic laws used to make the deduction.

The deductions of the local propagator are used not only for their simple explanatory capability, but as a source of data by the other two experts. Although the local propagator appears simple, it is actually quite complex because of the need for generality: It must deal with aribitrary measurements in arbitrary circuits. The gain from building a general propagator is great because it is the only part of Sophie III that needs to reason upon the measurements and circuit topology directly. This greatly simplifies the other experts because they need only refer to the propagator's deductions.

To describe the work done by the local propagator, we will present an example of how the propagations of a single measurement can lead to quite powerful conclusions. For our example, we will use the circuit shown in Figure 9. Suppose we measure the current through R_{13} and find it to be 1 mA. From the circuit schematic information we determine that the resistance of R_{13} is 100 Ω. Hence, from Ohm's law, the voltage across the base–emitter junction of Q_6 must be approximately .1 V. We know that a silicon transistor conducts no current when the base–emitter voltage is less than .6 V, so the current flowing in each of the transistor's terminals must be zero. Now consider node N_9, which connects to R_{13}, Q_6, and R_{16}. The current flowing out of R_{13} and Q_6 must be flowing into resistor R_{16}, so the current flowing through R_{16} is 1 mA. Thus, we can look up the resistance R_{16} and determine the voltage drop across it as well.

As we have just seen, a single measurement can lead to the calculation of a lot of information about other circuit values. The local propagator of Sophie III performs these deductions and keeps records of each deduction so that explanations can be generated and examined when troubleshooting. Although this type of reasoning, in the abstract, seems complicated, the schematic makes it wonderfully transparent. No expressions need to be propagated, just numbers or intervals; furthermore, the schematic acts as a means of controlling the deductions and recording their results.

As powerful as the local propagator is, it is still not adequate for tracking the troubleshooting of the student. This is particularly evident early in a debugging

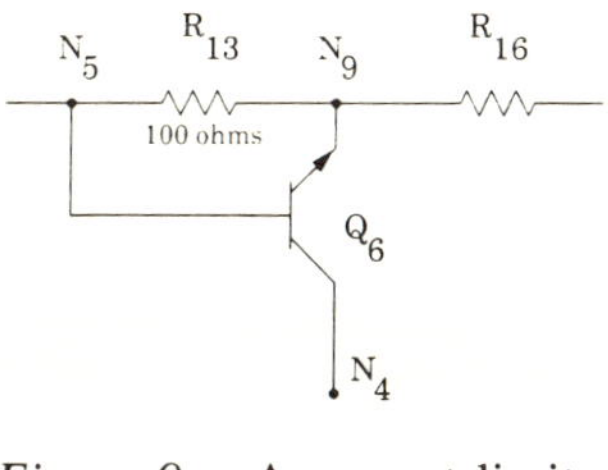

Figure 9. A current limiter.

scenario. The first measurement every debugger performs is to measure the output. This is probably the single most informative measurement he can make. But the propagator can do nothing with it. Only after a number of measurements have been made can enough values be propagated to derive useful information as to what can be wrong. Clearly the propagating scheme, by itself, is insufficient. But what is missing?

The propagator is missing all the attributes of "understanding how the circuit works." It does not understand the causality or the teleology of the circuit; in fact, it does not even have a hierarchical description of the functional components. It employs a myopic or local view of the circuit and thus fails to encompass the global mechanism through which the circuit functions. Its myopic view is only appropriate when the fault has been isolated to a module or group of components; it is entirely inappropriate for determining the meaning of initial measurements or, for that matter, of any measurements made at the functional block level.

When Sophie III was started, the existing theories about causal and teleological reasoning were insufficient to permit their inclusion in the propagator in a general way. This is why Sophie III is structured in levels. Levels 1 and 2 capture a summary of the *results* of such reasoning in the form of production rules and level 3 uses these to make deductions. de Kleer and Brown employed two principles to encode the circuit-specific knowledge. First, its form is severely restricted, making addition, modification, and explanation of higher-level knowledge easier. Second, the propagator is used as a way of simplifying the deductions that the higher-level knowledge needs to make. Consider a simplistic example stemming from a regulated power supply that had the rule "if output voltage is low, the regulator is shorted." The output voltage can be determined in a number of ways, one of which is by measuring the current through a resistor placed across the output terminals and using Ohm's law to determine the voltage across it. The rule tells us that if the voltage is directly measured to be low, the regulator must be shorted. If the current is measured to be low, then either the regulator is shorted or the load is open or high. By utilizing the propagator and using the assumptions of the propagation (in this example, that the resistance of the load hasn't changed) to modify the consequences of the higher-order rules, only one rule is needed to handle both of these circumstances.

Work on Sophie III proceeded along many other vectors as well, but the project was ceased for two reasons, one pragmatic and one intellectual. The pragmatic issue was that to make further progress a larger address space and a more powerful computer for running Lisp was needed; the intellectual issue was our increased awareness that we did not really know what it means to "understand" how a complex piece of equipment works. In particular, we did not know what mental models experts have of a given system's functioning, nor did we know how these models are learned, for they certainly aren't explicitly taught. In brief, we needed a much better representation language and calculus for exploring the mental models and qualitative reasoning of experts in electronics, especially in feedback systems.

VI. QUALITATIVE REASONING AND MENTAL MODELS

Understanding how experts construct and use mental models to cope with feedback in complex systems was harder than it at first appeared. While the experts we interviewed had very little difficulty in coming up with models to explain any given situation, getting a systematic framework that could determine ahead of time which model to use and how to apply it was extremely hard. It turns out that the models are specific to the particular situation or context in ways that are not captured in the models themselves. To be more concrete, a transistor can be viewed as an amplifier or as a switch, depending how it is connected to other components. An explanation that says "the current that comes in the base is amplified through the emitter" assumes that the transistor in question is on and is functioning in its linear range. While this may be a reasonable assumption for a particular working circuit through its normal range of operating conditions, it is not universally true. It seems that major problems are solved by an expert explainer in choosing the particular view of a component to use in a situation and that the inference that this decision is based on is not made explicit in the explanation. Without this additional information (about the assumptions under which a given view is valid), the student will not know when to apply the view (and likely won't be able to remember it anyway). We want to stress here that this is not an argument against a particular view, but that a mental model of how an overall device functions usually rests on assumptions that determine how to view components and their interactions at a given time. In particular, it argues against focusing on teaching the student models of an overall device in favor of teaching him how to build models of devices in the first place.

Because of the contextual dependency of models, de Kleer and Brown developed a calculus for constructing qualitative, causal models of devices rather than teaching the particular models of a given expert. Indeed, as demonstrated in the following example, even textbooks gloss the *implicit* context dependencies buried in their explanations. Since these assumptions remain implicit and yet are still idiosyncratic to the particular device, working in a particular context—that is, they are not just part of the background assumptions that necessarily underlie any explanation—these explanations are more like rationalizations. On hearing it, it seems to have explanatory force, but when one goes to reconstruct the explanation one suddenly discovers that there are many other possible rationalizations (i.e., "explanations") that also could be true *and* which predict quite different device behaviors.

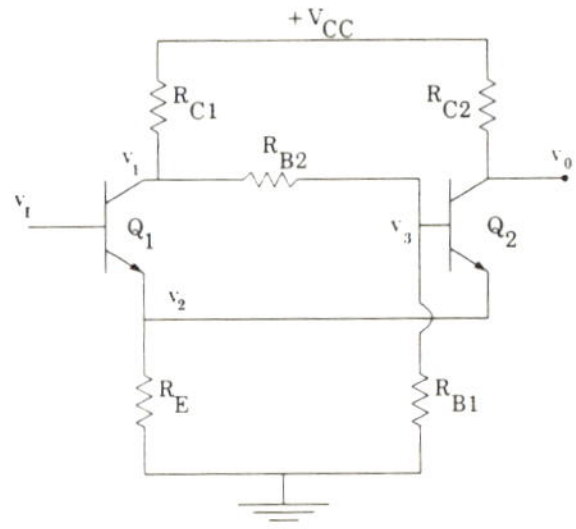

Figure 10. The Schmitt trigger.

Consider the following explanation of a Schmitt trigger (shown in Figure 10) taken from a textbook in electronics (Harris et al., 1966, p. 68):

> An increase in V_1 augments the forward bias on the emitter junction of the first transistor, thereby causing an incremental increase in the collector current, I_{C1}, of that transistor. Consequently both the collector-to-ground voltage V_1 of the first transistor, and the base-to-ground voltage of the second transistor v_3, decrease. The second transistor operates as an emitter follower which has an additional load resistor on the collector. Therefore, there is a decrease in the emitter-to-ground voltage V_2. This decrease in V_2 causes the forward bias at the emitter of the first transistor to increase even more than would occur as a consequence of the initial increase in V_1 alone. . . .

This ''explanation'' for the Schmitt trigger makes a number of unstated assumptions. Why does the v_I increment appear across Q_1 instead of R_E? Why does the voltage V_1 drop when Q_2's turning off should raise it? Why is the current through R_E contributed by Q_2's turning off more than the current by Q_1's turning on?

There are many values for the parameters for which the circuit cannot function at all. The above ''explanation'' is only a rationalization of the observed behavior (observed by actual measurements or stated in the textbook) and cannot necessarily be used to understand other circuits. This does not detract from the usefulness of the explanations; no explanation ever accounts for every detail of device behavior. But it does argue that in becoming experts, we become aware of the assumptions underlying such explanations so that we can reconstruct them when needed and generalize them to remain applicable when certain faults are introduced.

A. Mental Models That Perform Work

Explanations in our view turn on two categories of sequenced mental models. The first consists of a space of models that have a fixed grain size but in which each successive model contains fewer implicit assumptions about the context of the components of the device. For example, we might model a light switch on the wall by saying that when it is turned on, current flows, turning on the light. But this simple model assumes that there is a potential difference across the switch which need not be the case. For example, the switch might be in series with another switch or circuit breaker.

The embedding of implicit assumptions in the models is addressed by our fundamental principle of "no function in structure." See de Kleer and Brown (1983, 1984) for detailed discussions of this principle and how it relates to the issue of multiple mental models. It is perhaps worth noting here that one aspect of the "no function in structure" principle is easily misunderstood. The principle does not claim that the only kind of mental models possessed by experts, in contradistinction to novices, are those that satisfy the "no function in structure" principle. On the contrary, both experts and novices possess multiple models of a system, many, if not all, of which violate the ideal of "no function in structure." The difference between experts and novices, said simply, is that the models that experts possess will have implicit assumptions embedded in them but they will also have *explicit* assumptions wrapped around that model which determine when that particular model may be safely used. Thus, when troubleshooting certain kinds of faults, an expert will know immediately when to move to a more *pure* model that does not presuppose some aspect of the correctly functioning circuit which may now be in question because of some conjectured fault.

The second category of sequenced models is the standard one in which the grain size of the modeling components is changed. Thus, for example, at one level Kirchoff's law will suffice for modeling the flow of current around a wire. At a finer grain size, we might have to consider Maxwell's equations and the speed of propagation of electromagnetic perturbations. The grain size issue is governed by classwide assumptions, which function as the background assumptions against which the "no function in structure" principle is applied.

B. Envisioning vs. Envisionment

An often overlooked key to successful troubleshooting is the ability for the troubleshooter to move between structure and function. Understanding, actually, is more a *relationship* between the structure and function of a device than just a causal simulation of its overall function. It does little good to have a deep envisionment or causal mechanism of how a particular device functions if in fact that envisionment is not derivable from the structure of the device itself. This observation becomes most apparent when considering troubleshooting a fault that has changed the overall behavior of the device in fundamental ways. Faced with such a situation, the troubleshooter must be able to reason from the structure and potential perturbations to that structure to changes in the behavior or functioning of the device. That is, he must be able to invert the mapping between structure and function. Note that the crucial distinction between envisioning and envisionment lies at the heart of this issue. That is, envisioning is the *problem-solving process* that one goes through in constructing the relationship between structure and function. The envisioning process reasons over the device topology and constructs a class of possible causal mechanisms each resting on some set of assumptions. These causal mechanisms linked with their assumptions are called *envisionments*. Thus, the product of the envisioning is a set of envisionments.

It is important to recognize this distinction when considering the cognitive structures that we want the student to develop. Observing the simulation of the device must not only help the student to construct causal attributions to the events that transpire, but he must also be aware of the assumptions underlying why those events occurred as they did in the first place. Ideally, we would like the student to develop the skill of being able to construct a set of causal models from the device's structure, thus guaranteeing that he knows the mapping between the device's structure and its function under all boundary conditions.

C. Coping with Feedback and the Component Heuristic

Thus far, we have discussed the role of local propagation of measurements in Sophie III, mental models, implicit assumptions buried in those models, and the movement from structure to function. We have also mentioned that what often makes troubleshooting difficult is feedback and that since feedback is a functional property rather than a structural property of the circuit, just detecting its presence can be problematic. The key to recognizing and modeling feedback lies in combining the method of local propagation with a simple "causal" heuristic, a heuristic that helps propagate an input perturbation around the circuit. This heuristic is called the *component heuristic*. Metaphorically, it says that if there is a "push" or "pull" on one side of a component and nothing else is yet known to be acting on the other side of the component, then the component responds as if the unknown actions are negligible. See de Kleer and Brown (1984) for discussion of the rationale and use of this heuristic.

This heuristic when coupled with a modified version of Sophie III's propagation scheme provides a simple way to reason qualitatively about circuits that have feedback. Analyzing circuits with feedback usually requires solving a set of simultaneous equations. The simultaneity comes from the cycle of information flow in the feedback system and the inherent interdependencies that determine state parameters of the circuit. The key role for the component heuristic is to break the inherent dependency cycle by introducing an assumption of primacy and then to propagate the results of that assumption around the circuit until the assumption folds back on itself—thereby detecting a feedback cycle.

Instead of providing a formal analysis of this scheme which can be reviewed in de Kleer and Brown (1984, 1986), we will illustrate its application to the classic example of a voltage divider—one of the simplest circuits whose behavioral analysis requires the use of a set of simultaneous equations. Let us consider the voltage divider comprised of two resistors shown in Figure 11.

If the voltage across nodes N_1 and N_3 experiences a small positive increment what happens to the voltage at node N_2? To determine the answer, we need to apply a qualitative version of Ohm's law which says that if the potential difference across R_1 increases, then the current through it will also increase. How

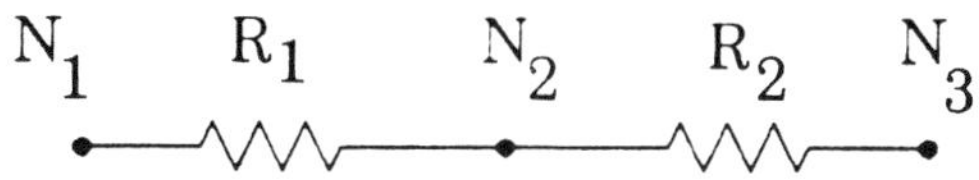

Figure 11. A voltage divider.

ever, before this law can be applied, the voltage difference across R_1, not just the voltage at one of its nodes, must be known. Consequently, the propagator must come to a halt. Not enough information is *locally* known at this instant in order to proceed. Now, let's apply the component heuristic to R_1. The heuristic says that if you don't yet know anything about what is happening at N_2, *assume* that it is negligible. With that assumption in place, the propagator can now proceed, since an increment of the voltage at N_1 implies an increment of the voltage difference across R_1 via the assumption. Now, Ohm's law can apply, deducing that there is an increase of current through R_1.

Similarly, the increment of current flowing out of R_1 goes into R_2, which implies, via another application of the qualitative version of Ohm's law, that the voltage difference across R_2 also increases. This leads one to conclude that the voltage at N_2 rises. However, we had earlier assumed, with the first application of the component heuristic to R_1, that the voltage change at N_2 was negligible. Thus, we have reached a contradiction. This is exactly what we were looking for. The assumption enabled us to make inferences that eventually (two steps later) fold back on itself. This foldback is the crucial step in detecting feedback which, in this case, is negative feedback.

Note that while this is a simple case, it is nevertheless tricky, since what makes analyzing a voltage divider ordinarily require solving two simultaneous equations is precisely the negative feedback that emerges from the way that these two resistors interact. A similar, but much more complex, analysis works for the Schmitt trigger discussed earlier. For the Schmitt trigger, the component heuristic coupled to the propagator not only identifies the feedback, but also identifies many of the implicit assumptions that were embedded in that explanation.

In summary, in the last few years de Kleer has managed to formalize a qualitative calculus and a simple propagation scheme for "solving" qualitative constraints that model circuit components and electronic laws and to define a set of heuristics that can be used to the identify the assumptions and feedback paths in any analog circuit. (See de Kleer and Brown, 1984, for a full discussion of this calculus.) Interestingly, applying the calculus via the propagation mechanism is straightforward and does not involve the solution of any equations! Instead, one can just propagate qualitative increments/decrements around the circuit along with the appropriate assumptions and through this process construct the set of all possible, causal models consistent with the circuit schematic. This, in essence, is the envisioning algorithm, an algorithm that has been used extensively in the last few years for constructing electronic expert reasoners. (See de Kleer, 1984, and de Kleer and Brown, 1986, for more discussion of this algorithm.) Although we have yet to apply these findings in an educational setting, we believe they will form the foundation for a new way to teach students how to qualitatively understand electronics and how to construct causal mental models of how systems work by examining their structure.

VIII. LOW-ROAD APPLICATIONS OF HIGH-ROAD TECHNOLOGY

One of the patterns that we have noticed in years of working on training problems is that the high-road approach of developing sound theories about a subject domain and computationally expensive systems that exploit those theories often results in spinoffs that have more immediate low-road application. In this section we will describe a spinoff of the Sophie technology.

The bulk of computational resources needed to run Sophie I went into the circuit simulation and the hypothesis evaluation programs. Most of the time, these parts of the system were not being used; the simulator was needed only when the student changed a control or inserted a fault, and hypothesis evaluation was needed only when the student ventured a hypothesis. As an experiment in low-cost delivery, we built a distributed instructional system using Apple IIs as workstations coupled to a fast mainframe computer containing Sophie's general-purpose circuit simulator and AI inference capabilities. The computational capabilities of the mainframe were used to automatically generate voltage tables characterizing the behavior of the circuit for a *prespecified set* of faults as well as for its normal working behavior. These automatically generated tables were then stored locally on the microcomputers. The microcomputer was made to "look" like a workbench by attaching to it a bitpad tablet on which color photographs of the actual printed circuit boards or schematics were laid.

In the micro, a model of the photograph was constructed that mapped the bitpad coordinates to the electrically significant terminals and junctions. This made it possible to use the tablet's pen as a simulated voltage probe. Students could make any measurement they wished simply by using the tablet's pen to touch the appropriate terminal on the photograph. The use of the bitpad and photograph had the additional benefit of introducing another representation of the circuit (how the circuit actually looks on a circuit board) and gave the student practice mapping between this representation and the schematic.

The micro not only simulates the broken and working device, but, crucially, keeps track of all of the student's measurements and attempted component replacements. It can also accurately record the time at which each measurement was made. At any time, the student can review the measurements he has taken and ask for prestored help.

After he finishes his troubleshooting exercises, the history of his interactions is

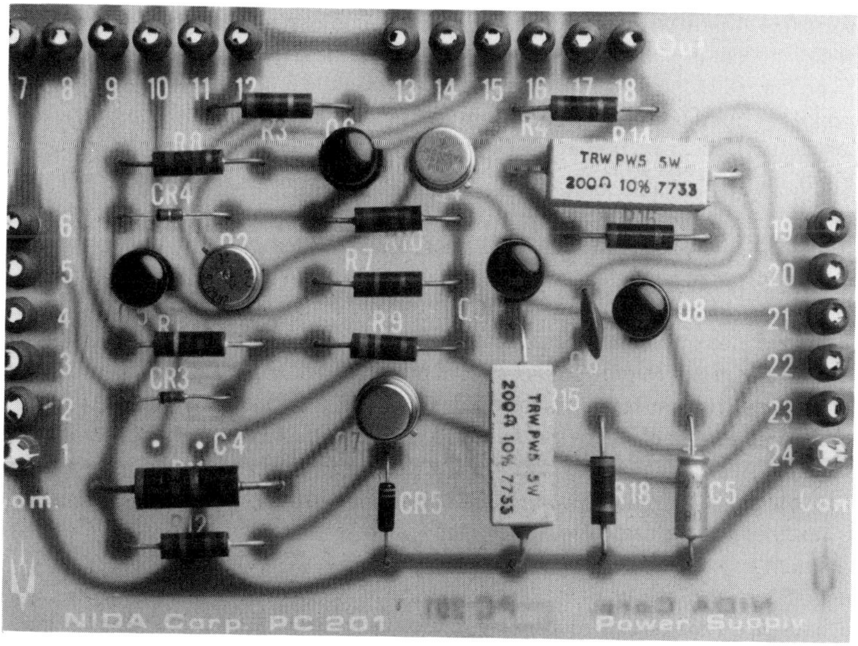

Figure 12. Photograph of the PC board of a power supply modeled by Sophie IV.

transferred to the mainframe and critiqued, off-line, by the Sophie coach. The coach determines what set of faults the student should try next day and automatically creates the data base for the micro's table-driven simulator.

One of the nice things about this scheme, and something that is unlike making measurements in the real world, is that an audit trail of the particular measurements is kept automatically. This record can then be reviewed at any point, an important capability for several reasons. First, the Sophie lab often tries to cause the student troubleshooter to think about the circuit by having him troubleshoot with the minimal number of measurements. This constraint encourages the student to reason as far as he can about the consequences of any particular measurement. Without the audit trail, he has to write down each measurement and its value to help him avoid making a redundant measurement. The history list relieves him of this burden while demonstrating the usefulness of reviewing his problem-solving steps. Second, the history list captures all the noteworthy steps in the student's troubleshooting process, giving the computer-based coach a useful account to critique.

To encourage the student further to think through the entailments of each measurement, we provided additional power tools to help him propagate the consequences of any measurement. For example, we simulated an RPN calculator that allows the student to load any measurement from the history list directly into the calculator. However, unlike most calculators, this one does not deal just with numbers, but with the particular units of a given measurement that had been made, e.g., voltage (volts), current (amperes), resistance (ohms). As a result of these capabilities, the calculator supports an operationalization of Ohm's Law such that, for example, when the student multiplies a particular current times a resistance, the calculator would return the answer in terms of volts.[7]

IX. CONCLUSION

A. A Futuristic Learning Scenario

Advances in computational power, display technologies, and storage media make possible a new kind of learning scenario, one that allows use of the ideas discussed in this paper to address a growing educational problem—the need for more effective on-the-job training. Ironically, a large part of the need for on-job-site training stems from the rate of change of today's technology. It no longer makes sense to train technicians at the beginning of their careers, expecting that this training will be sufficient learning for their entire career. At the pace that technology is changing, training a technician to understand today's technology, even with the deeper conceptual models talked about in this paper, may not be adequate preparation to understand the equipment they will see five years hence. The very paradigms underlying the next generation of technology are apt to be different from those underlying today's technology. In addition to being able to continually update a technician's knowledge, on-job-site learning also reduces the time that trainees are removed from the field and facilitates situated learning much akin to the apprenticeship learning of the past eras.

The key to this new style of learning is the notion of intelligent mentors that can act as job performance aids, not only to help the trainee perform his task, but more importantly to transform the particular problems that the trainee encounters in the field into significant learning events. This vision is now feasible because of the advances in technology, in expert systems, in coaching systems, and in the cognitive and social sciences:

Technology. Even the most conservative extrapolation of today's technological developments suggest that within a small number of years, we will have lap-sized computers equal in power to our most powerful AI/Lisp machines of today. Displays of the quality to allow high-grade animation will be here, as will portable laser disks that are capable of storing the massive amounts of information underlying today's technical documentation systems and video sequences. Thus, tomorrow's technician can have at his fingertips computational power and knowledge bases akin to those in a few research laboratories today.

Expert systems. The recent advances of the Guidon System (Clancey, 1982; Richer and Clancey, 1984) together with the research into qualitative reasoning on mechanistic systems (de Kleer and Brown, 1981, 1983, 1984, 1986) indicate that we can now construct moderately robust expert systems that perform serious work in styles that are both strategically and tactically similar to humans. Such systems, extended to incorporate more psychologically oriented work on mechanistic mental models, provide the basis for implementing serious apprenticeship learning. The user or apprentice can receive advice about how to handle particular situations in a way that enriches his understanding of the phenomena and helps him develop more robust mental models or conceptualizations of its underlying mechanisms. The apprentice in this scenario has an advantage over his medieval counterpart in that he can use the technology to observe aspects of phenomena from different perspectives and can be shown animated, conceptual entities "flowing" through the structure of the machine. Often the entities out of which useful causal models are built are unobservable because they are simply abstractions. Some examples are the flow of electrical current, the flow of electrons, propagation of pressure, propagation of changes of energy, force field lines, and entropy. Being able to observe how these conceptual entities move through the system and interact with each other to produce the observed behavior opens up a new era in situated learning, learning that is not only guaranteed to be relevant since it addresses the problem at hand—the apprentice's problem, and a real problem—but also that helps him to formulate more robust conceptual models of how the structure of the device enables its function.

B. Social Structure of Ongoing Learning

One final comment might be in order. Much of this chapter has stressed the use of advanced technology and theories of how the mind represents and uses knowledge to structure learning environments. We want to close on a low-technology note concerning how the sociology of learning and the social context or infrastructure that enables the spread of knowledge can provide tremendous leverage for technical training. Recent anthropological studies (Orr, 1986) indicate that a prevalent form of technical communication between peers in the field is in the form of telling narratives (stories) about particularly difficult repairs. The "war story" so often heard over lunch or coffee is actually serving the major role of transmitting community knowledge about the equipment and new failure modes. What makes this interesting from a training perspective is that the stories appear to have a significant structure which is expected by the community. War stories are accounts of how the equipment in question works (or fails) in terms of models that account for the particular symptoms observed. These stories include all the contextual details necessary to support the assertions about the model, and these details aid new technicians in developing and refining their model of the machine. The import of this for training is that the skill of understanding and telling stories can be taught as part of the initial training. Thus, the student will come out with a learning skill that will continue to serve him throughout his life.

ACKNOWLEDGMENTS

We gratefully acknowledge the following people who contributed to the work described in this paper: Johan de Kleer for Sophie III and mental models; Alan Bell for Sophie I; Richard Rubinstein and Ned Benhaim for Sophie II's articulate expert; Daryle Lewis for the Apple II version and Austin Henderson for Alto version of distributed Sophie; and Allan Collins for discussions about reciprocal teaching. We would also like to express our gratitiude to Carlin DeCato, Susan Newman, and Austin Henderson for assistance in editing this paper.

NOTES

1. Indeed, the history of science is the chronicle of humankind's discovering increasingly more useful eyeglasses. The length of time between changes in eyeglasses—i.e., between paradigm shifts—is testimonial to the difficulty of breaking conceptual set without outside help that points out the relevant inconsistencies.
2. Interestingly, these mechanisms also function as the primary explanatory mechanisms that comprise causal understanding within the conceptual system. An example of a mechanism is the pressure waves within fluids that carry the information of pressure changes.
3. Note that a given set of measurements might not have isolated the potential faults to just one. In such a case, the student could have jumped prematurely to a conclusion about what is wrong. In such cases it is important that if his guess is wrong, that he learn that it is wrong but still logically consistent with the values of all the measurements thus far made. Similarly, it is important for the system to identify when the student is jumping to a conclusion, even if it is right. Jumping to conclusions can be an optimal strategy when the alternatives have low probability of existing.
4. A teleological model is one in which intended purposes are described.
5. Also see Collins and Brown (1986) for a further description of this and other related work.
6. We discussed earlier that a person's understanding determines what they see. Figure 3 provides an example. To people who don't know any electronics, Figure 3 is a collection of lines. To people with some familiarity with electronic notation, it is a piece of a circuit made up of transistors and resistors. To an expert troubleshooter, it is a collection of active elements that produces a constant current.
7. Schwartz (1982) has developed the use of a similar tool in word problems.

SUGGESTED FURTHER READING

Qualitative Reasoning About Physical Systems, edited by Daniel G. Bobrow. Cambridge, MA: MIT Press, 1985.

This book contains state-of-the-art papers on qualitative reasoning about physical systems. Some of the papers described here discuss qualitative reasoning from the perspective of a device-centered ontology. Others take a process-centered ontology as their foundation and discuss formalisms and theories about how to represent the reasonings inherent in naive physics. Of particular relevance is the paper by Forbus on qualitative process theory and his earlier work on a feedback minilab for students (Forbus, 1984). The last set of papers shows how one can construct troubleshooting systems that reason from first principles as opposed to the more experientially oriented expert systems approaches, such as MYCIN.

Intelligent Tutoring Systems, edited by Derek Sleeman and John Seely Brown. New York: Academic Press 1982.

This book contains many of the classic papers in the new area of intelligent tutoring systems. The papers range from cognitive and psychological studies to descriptions of many classic intelligent tutoring systems.

Mental Models, edited by Dedre Gentner and Albert L. Stevens. Hillsdale, NJ: Erlbaum Associates, 1983.

This book contains one of the best collections of research papers on mental models. Many of the issues discussed in this paper are touched on by various contributors to this book. Although the focus of most of the papers in this book are not primarily educational, it's not difficult to translate the issues discussed here into ones that directly impact training and education.

REFERENCES

Brown, A.L., and Palincsar, A.S. 1986. Reciprocal teaching of comprehension strategies: a natural history of one program of enhancing learning. In J.B. Borkowski and J.D. Day (Eds.), *Intelligence and Cognition in Special Children: Comparative Studies of Giftedness, Mental Retardation, and Learning Disabilities*. Norwood, NJ: Ablex.

Brown, J.S. and Burton, R.R. 1975. Multiple representations of knowledge for tutorial reasoning. In D.G. Bobrow and A. Collins (Eds.), *Representation and Understanding*. New York: Academic Press.

Brown, J.S., Burton, R.R., and Bell, A.G. 1975. SOPHIE: a step towards a reactive learning environment. *International Journal of Man Machine Studies 7*:675–696.

Brown, J.S., Burton, R.R., and de Kleer, J. 1982. Pedagogical, natural language and knowledge engineering techniques in SOPHIE I, II, III. In D. Sleeman and J.S. Brown (Eds.), *Intelligent Tutoring Systems*. New York: Academic Press.

Brown, J.S., Rubinstein, R., and Burton, R.R. 1976. *Reactive Learning Environment for Computer Assisted Electronics Instruction*. Cambridge, MA: Bolt Beranek and Newman, Inc., Report No. 3314.

Burton, R.R. 1982. Diagnosing bugs in a simple procedural skill. In D. Sleeman, and J.S. Brown (Eds.), *Intelligent Tutoring Systems*. New York: Academic Press.

Burton, R.R. and Brown, J.S. 1978. Diagnostic models for procedural bugs in basic mathematical skills. *Cognitive Science 2*:155–192.

Burton, R.R. and Brown, J.S. 1979. Toward a natural-language capability for computer-assisted instruction. In H. O'Neil (Ed.), *Procedures For Instructional Systems Development*. New York: Academic Press.

Burton, R.R. and Brown, J.S. 1982. An investigation of computer coaching for informal learning activities. In D. Sleeman and J.S. Brown (Eds.), *Intelligent Tutoring Systems*. New York: Academic Press.

Clancey, W.J. 1982. Tutoring rules for guiding a case method dialogue. In S. Derek and J.S. Brown (Eds.), *Intelligent Tutoring Systems*. New York: Academic Press.

Clement, J. 1983. A conceptual model discussed by Galilleo and used intuitively by physics students. In D. Gentner and A.L. Stevens (Eds.), *Mental Models*. Hillsdale, NJ: Erlbaum Associates.

Collins, A. and Brown, J.S. In press. The computer as a tool for learning through reflection. In H. Mandl and A. Lesgold (Eds.), *Learning Issues for Intelligent Tutoring Systems*. New York: Springer.

de Kleer, J. 1979. *Causal and Teleological Reasoning in Circuit Recognition.* Cambridge, MA: MIT Technical Report No. 529.

de Kleer, J. 1984. How circuits work. *Artificial Intelligence 24:*205–280.

de Kleer, J. and Brown, J.S. 1981. Mental models of physical mechanisms and their acquisition. In J.R. Anderson (Ed.), *Cognitive Skills and Their Acquisition.* Hillsdale, NJ: Erlbaum Associates.

de Kleer, J. and Brown, J.S. 1983. Assumptions and ambiguities. In D. Gentner and A.L. Stevens (Eds.), *Mental Models.* Hillsdale, NJ: Erlbaum Associates.

de Kleer, J. and Brown, J.S. 1984. A qualitative physics based on confluences. *Artificial Intelligence* 24:7–83.

de Kleer, J. and Brown, J.S. 1986. Theories of causal ordering. *Artificial Intelligence 28.*

di Sessa, A. 1983. Phenomenology and the evolution of intuition. In D. Gentner and A.L. Stevens (Eds.), *Mental Models.* Hillsdale, NJ: Erlbaum Associates.

di Sessa, A. 1982. Unlearning Aristotelian physics: a study of knowledge-based learning. *Cognitive Science 6:*37–75.

Forbus, K. 1984. An interactive laboratory for teaching control system concepts. BBN Laboratories. Report No. 5511, Cambridge, MA.

Genter, D. and Gentner, D.R. 1983. Flowing waters or teeming crowds: mental models of electricity. In D. Gentner and A.L. Stevens (Eds.), *Mental Models.* Hillsdale, NJ: Erlbaum Associates.

Harris, J.H., Gray, P.E. and Searle, C.L. 1966. *Digital Transistor Circuits.* New York: Wiley.

McCloskey, D. 1983. Teaching thinking: a cognitive-behavioral perspective. In S.F. Chipman, J.W. Segal and R. Glaser (Eds.), *Thinking and Learning Skills,* Vol. 2. Hillsdale, NJ: Erlbaum Associates.

Meichenbaum, M. 1985. Naive theories of motion. In D. Gentner and A.L. Stevens (Eds.), *Mental Models.* Hillsdale, NJ: Erlbaum Associates.

Orr, J. 1986. Talking about machines: social aspects of expertise. Xerox PARC, ISL working paper, Palo Alto, CA.

Palincsar, A.S. and Brown, A.L. 1984. Reciprocal teaching of comprehension-fostering and monitoring activities. *Cognition and Instruction 1:*117–175.

Richer, M. and Clancey, W.J. 1985. GUIDON–WATCH: a graphic interface for viewing a knowledge-based system. *IEEE Computer Graphics and Applications 11:*51–64.

Schwartz, J.L. 1982. The semantic calculator. *Classroom Computer News 2*(4):22–24.

VanLehn, K. and Brown, J.S. 1980. Planning nets: a representation for formalizing analogies and semantic models for procedural skills. In R.W. Snow, P.A. Fredico, and W.E. Montagues (Eds.), *Aptitude Learning and Instruction,* Vol. 2: *Cognitive Process Analyses of Learning and Problem-Solving.* Hillsdale, NJ: Erlbaum Associates.

White, B.Y. 1984. Designing computer activities to help physics students understand Newton's laws of motion. *Cognition and Instruction 1*:69–108.

White, B.Y. and Frederickson, J.R. 1986. Progressions of qualitative models as a foundation for intelligent learning environments. BBN Laboratories, Report No. 6277, Cambridge, MA.

Williams, M. and Hollan, J. 1981. The process of retrieval from very long term memory. *Cognitive Science 5:*87–119.

Williams, M., Hollan, J., and Stevens, A. 1981. An overview of STEAMER: an advanced computer assisted instructional system for propulsion engineering. *Behavior Research Methods and Instrumentation 13*(2):85–90.

Wiser, M. and Carey, S. 1983. When heat and temperature were one. In D. Gentner and A.L. Stevens (Eds.), *Mental Models.* Hillsdale, NJ: Erlbaum Associates.

DEVELOPMENT AND EVALUATION OF SIMULATION-ORIENTED COMPUTER-BASED INSTRUCTION FOR DIAGNOSTIC TRAINING

William B. Johnson

ABSTRACT

For over 10 years the author and his colleagues have been engaged in an evolving research and development program on computer-based diagnostic training. Starting with basic research on human problem solving and continuing to practical applications in numerous technical environments, the research has contributed significantly to defining the characteristics of effective computer-based diagnostic training. Each of the research steps has left as operational computer-based training system in place. This chapter describes the evolution of this research, as well as the resulting training systems that are currently in use in a variety of industrial and military domains.

Advances in Man-Machine Systems Research, vol. 3, pages 99–127.

ISBN: 0-89232-659-X

I. INTRODUCTION

For several decades, the ''new'' technology of computer-based instruction (CBI) has been heralded as a revolution in training (Alessi and Trollip, 1985). However, it has not been widely adopted in industrial environments. It may be that the basic researchers in education, psychology, computer science, and CBI-related fields find laboratory results that are not attuned to the reality of training development and delivery in industrial and/or military environments. Perhaps the research is published in journals read by other researchers rather than by training practitioners. While educational researchers search for the ''cutting edge,'' hardware vendors take the message to the trainers. All too often the vendors overemphasize the computer systems, videodisks, and other devices while ignoring the all-so-important process of training development, delivery, and evaluation.

This chapter describes a continuous 10-year research and development effort on computer-based training for diagnostic-related instruction that, in an organized way, began as basic research and continued into applied instructional development for industrial and military applications. The emphasis is on simulation-oriented computer-based instruction (SOCBI). The chapter will highlight the factors that have caused SOCBI, unlike many other training research vehicles, to be accepted as a thriving part of the instructional environments in which it has been introduced. These factors include the following:

- An attempt to understand human problem solving through basic research on diagnostic behavior
- An initial commitment that the simulation designed for research must also have immediate and continuing application
- Ongoing applications-oriented field evaluations with subsequent software modifications
- Continuous involvement of technical subject matter experts during conceptualization, development, implementation, and evaluation
- Controlled experiment evaluation involving assessment of learning transfer from CBI to operational equipment

II. SIMULATION-ORIENTED COMPUTER-BASED INSTRUCTION

A. SOCBI Defined

SOCBI is a type of computer-based instruction that permits the learner to engage in the cognitive process of data collection, hypothesis formulation, and testing which is representative of real-world technical troubleshooting. The term *simulation-oriented* is used because the training provides a simulated problem-solving scenario, although the program is not necessarily driven by the engineering models which characterize traditional simulation. For the most part, SOCBI does not present a dynamic system but merely different "snapshots" of the system in various failure or normal states. The various example systems described throughout this chapter serve to clarify the definition of SOCBI.

B. The Evolution of SOCBI

The SOCBI evolution with which this author has been involved began at the University of Illinois in 1976 and has continued to the present. It has developed in stages through a variety of computer simulations for diagnostic training in numerous instructional environments. Each stage has been marked by experiments which served to identify desirable as well as undesirable characteristics of the existing SOCBI version. The term *evolution* is used since the present simulations have evolved from an abstract problem-solving game, without a particular context, to very concrete representations of specific technical systems in such domains as aircraft mechanics (Johnson, 1980), ship operations (Rouse, 1982; Su, 1985), Army communications systems (Johnson and Fath, 1983, 1985), and, most recently, nuclear power plants (Johnson, Maddox, Rouse, and Kiel, 1985; Maddox, Johnson, and Frey, 1986). The evolution is characterized by numerous experimental evaluations that have been widely reported (Johnson, 1980; Rouse and Hunt, 1984). The remainder of this paper describes the major stages of the evolution.

C. Understanding Troubleshooting

The roots of diagnostic-related training, for this author, predate the aforementioned basic research on human problem solving.[1] Prior to that time a considerable experience base as a technical trainer in the environments of auto mechanics and aviation mechanics, as well as certification in both domains, indicated that technical competence is best demonstrated during troubleshooting activity. Richardson and his colleagues (1985) appropriately called troubleshooting the "heart" of maintenance. During a diagnostic task, the competent technician must integrate subsystem theoretical knowledge, practical operational experience, proper use and interpretation of instrumentation and test equipment, and a reasonable and logical approach to problem solution. In most cases this integration comes from experience, in training or on the job.

It is often the case that troubleshooting is learned on the job. However, on-the-job (OJT) diagnostic training is inefficient, ineffective, often unsafe, and simply not the preferred training method. In spite of these facts, OJT is probably the most widely accepted diagnostic training method. The cliché "learning from mistakes" clearly applies to OJT diagnostic training. The diagnostic training research described here has developed and tested appropriate methods to permit the troubleshooter to effectively "learn by mistakes" in the safe environment of computer simulation.

D. Troubleshooting by Application of Structural Knowledge (TASK)

The first problem-solving simulation reported here was developed for the National Aeronautics and Space Administration (NASA). The simulation, shown in Figure 1, was designed as a mechanism to explore human problem-solving performance. The diagnostic process for solving TASK problems requires that the troubleshooter find the one failed component in a randomly generated network. Each component in Figure 1 produces an acceptable (1) or unacceptable (0) output. All outputs emanating from a component carry the value produced by that component. In order for a component to have an acceptable output the component and all input components must be acceptable. By testing between components, as shown on the left column of Figure 1, the troubleshooter can isolate the failure within the system.

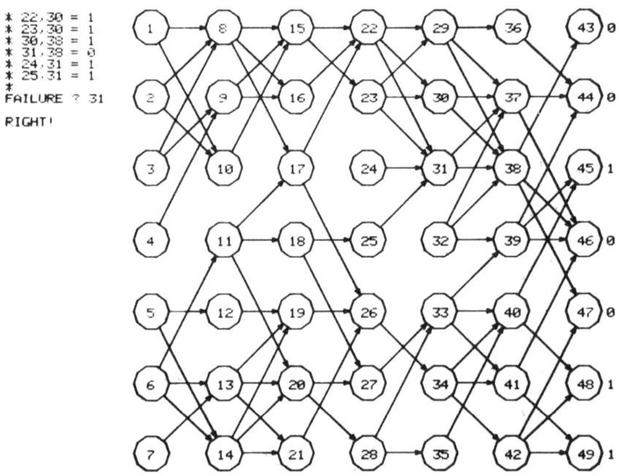

Figure 1. Example of TASK display.

TASK has been implemented on many computer systems and has had numerous variations, including network size, with 9, 16, 25, 36, or 49 components; touch-panel input; automatic computer aiding and performance feedback; and feedback loops within the component network.

Rouse and Hunt (1984) describe a series of five experiments, conducted between 1976 and 1978, that used TASK to study human problem solving. Results showed that the simulation not only served as a method for studying human problem solving, but also had the potential to provide training that could transfer to a wide range of real equipment. Troubleshooting appears to involve a set of generic skills combined with context specific technical knowledge. These early experiments explored the nature of these generic skills. However, it soon became clear that the study of problem solving and training for problem solving could not exist solely in an environment without context.

F Framework for Aiding the Understanding of Logical Troubleshooting (FAULT)

While TASK was a tool for basic research on human problem solving and showed potential for diagnostic training, there was a perceived need for a simulation more representative of "real-world" problem solving. Such a simulation had to embody the general problem-solving situations of TASK but offer the user the kind of diagnostic choices that would be available on real equipment. Therefore, the research team set out to characterize the necessary information for diagnostic decision making on real hardware. In addition, the team wanted to offer on-line information that would enhance training potential.

First, and perhaps foremost, it seemed that a troubleshooter must have the capability to collect information about the technical system. This would include reports from operators, operational status indications, and readings from appropriate instruments. The troubleshooter must have the capability to observe the status of equipment as it is installed and operating (or not) within the system. The troubleshooter must also have the option of removing components from the system for independent analysis.

From a training perspective, it was necessary that the simulation teach the troubleshooter about the functional relationships among the components in the system and, at least, basic operational definitions of each component. The work with TASK suggested that logical problem solving can be taught. Therefore, the simulation had to guide the learner to avoid inferential errors and capitalize on all the information gained during specific diagnostic tests. The first context-specific simulation was developed with these basic requirements.

FAULT emerged as a context-specific problem-solving simulation (Hunt, 1979; Hunt and Rouse, 1981). The program was developed to be adaptable to a variety of technical data bases. A highly evolved version of FAULT will be described here. Section III will show how some of the present capabilities of FAULT were conceptualized and finally integrated.

FAULT involves the use of a hard-copy schematic of a specific technical system (see Figure 2). The troubleshooter interacts with the hard-copy schematic via the CRT display shown in Figure 3. Depending on the technical environment, the troubleshooter is instructed to solve the problem with a minimum time and/or dollar expenditure. For example, the auto mechanic would want to minimize the cost of the repair, while the nuclear power plant technician would strive to minimize the threat of plant downtime. In addition, the troubleshooter is instructed to minimize redundant, unnecessary, or illogical tests. Such tests are considered to be errors.

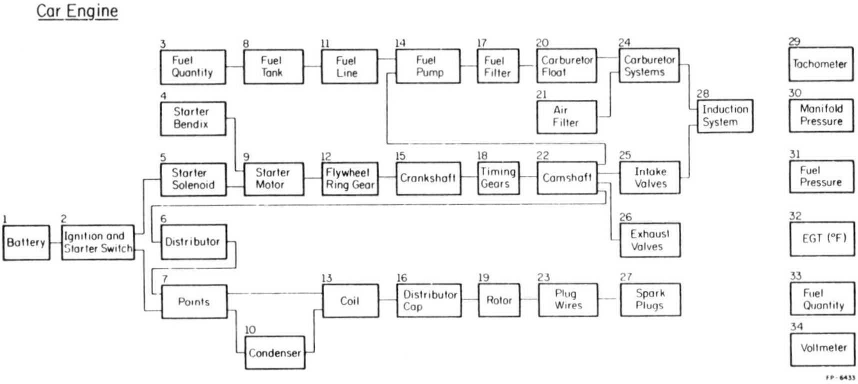

Figure 2. Example of FAULT schematic.

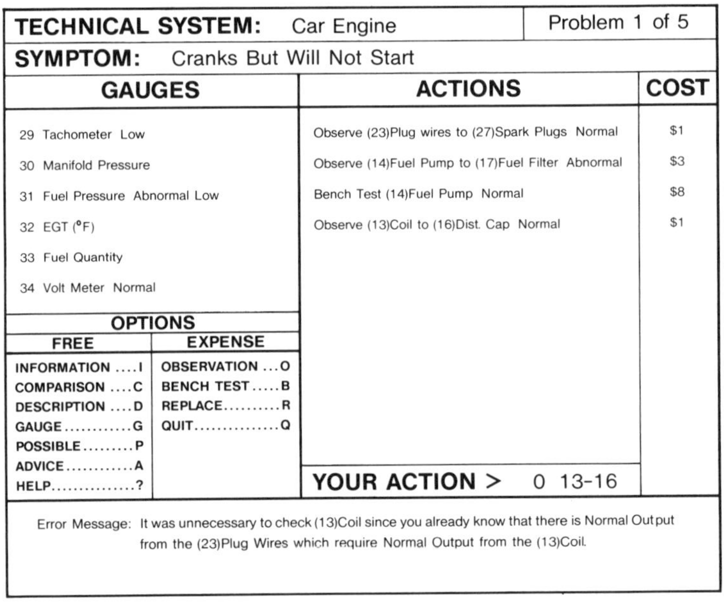

Figure 3. Example of FAULT display.

The lower left quadrant of Figure 3 divides options into columns labeled "Free" or "Expense." The "Free" options are as follows:

1. *Information*—learn how to perform various tests
2. *Comparison*—compare costs or time associated with tests or replacement
3. *Description*—learn functional description of each system component
4. *Gauge*—obtain a gauge reading

Table 1. Example Errors and Messages

Error	*Message*
Fixing part x when previous actions showed another part to be failed	This action was unnecessary. You have tested another part (name and #) and know that it was in need of repair.
Fixing part x that has no functional or logical connection to part y known to have an abnormal output.	This action was unnecessary. The output of x (name and #) does not reach y (name and #) which you know has an abnormal output.
Testing part x when all inputs are normal and output is abnormal.	This action was unnecessary. You know that x (name and #) has normal inputs and abnormal outputs. It must be abnormal.
Repairing part x after it tested normal.	This action was unnecessary. You have established that x (name and #) is normal. It did not have to be repaired.

5. *Possible*—list the "possible" set of failed components based on previous troubleshooting actions
6. *Advice*—learn the most logical tests to perform based on previous actions
7. *Help*—advice related to how to use the computer simulation

The "Expense" options include the following:

1. *Observation*—results of testing the output from one component to the next
2. *Bench test*—results of removing the component from the system and performing independent conclusive tests
3. *Replace*—replace the component
4. *Quit*—stop the program and receive the answer

1. FAULT Feedback

Feedback is provided throughout the problem, when errors such as those shown in Table 1 are made, and at problem termination in summary format. During a problem, FAULT will tell the student *when* an error is made and *why* the action was an error. The message at the bottom of Figure 3 shows an additional example of the type of feedback provided by FAULT. Table 1 shows examples from the 21 categories of problem-solving error and shows the feedback that is provided.

The FAULT performance summary is quite straightforward and will vary based on training specifications and computer system capability. Minimally, the feedback will acknowledge that the answer is correct and provide a summary of the number of actions performed, the number of errors related to diagnostic actions and part replacements, and the total time and associated costs for problem completion. On multiuser systems, data related to total number of previous simulation users and their average performance are also reported in the summary.

2. *Hierarchical FAULT*

As FAULT was used for more experiments and demonstrated across different technical disciplines, it became apparent that, at times, the program forced the user to replace rather than repair system components. Since the simulation was meant to offer the same cognitive diagnostic alternatives as the real world, a hierarchical display was developed. The hierarchical version of FAULT permits the user to troubleshoot at several system levels; for example, the problem can terminate by locating only the system (fuel–air), or the system component (fuel pump), or even a small subpart (fuel pump check valve) of one of the components.

III. INITIAL SOCBI RESEARCH AND DEVELOPMENT

The evolutionary development of TASK and FAULT described above was a result of developing, using, and evaluating SOCBI in numerous instructional environments. This section will consider each environment.

A. Aviation Maintenance Training

The first versions of FAULT were developed for aviation maintenance training at the University of Illinois under the sponsorship of the Army Research Institute (ARI) for the Behavioral and Social Sciences. The University of Illinois Institute of Aviation maintains a technical training program that prepares graduates for certification as airframe and powerplant mechanics by the Federal Aviation Administration. Between 1979 and 1981, five experiments were conducted using FAULT and TASK for aircraft mechanics students. Nearly 200 students participated in these experiments.

The first three experiments focused on using different versions of TASK to prepare students to use the FAULT simulation and on using different versions of FAULT to transfer students to FAULT simulations in unfamiliar technical domains (Hunt, 1979; Hunt and Rouse, 1981, 1984; Rouse and Hunt, 1984). These experiments suggested that computer-based aiding during training had the potential to improve diagnostic job performance. They also suggested that advanced technicians respond more favorably to the use of generic context-free simulations than do novice trainees. Advanced technicians were better able to transfer from generic training to context-specific training than were novice trainees. A second significant result of these early studies suggested that technical trainees can better develop generic problem-solving skills using multiple context-specific simulations than using a single context-free problem-solving simulation.

The fourth and fifth experiments were designed to compare SOCBI to traditional instruction (video-based) and to compare TASK with FAULT in its ability to prepare students for live equipment troubleshooting. The fourth experiment involved 36 advanced aviation maintenance trainees enrolled in the final course prior to FAA certification (Johnson, 1980; Johnson and Rouse, 1981a). The fifth experiment took place one year later and involved 22 advanced trainees (Johnson and Rouse, 1981a,b; Rouse and Hunt, 1984)

The results of the fourth experiment favored traditional instruction slightly, apparently due to the fact that the computer-based training resulted in procedural errors during real-equipment troubleshooting. It appeared that the SOCBI groups knew what they wanted to do but did not know precisely how to perform certain tests.

The findings from experiment four lead the researchers to two conclusions that affected subsequent SOCBI development, including immediate modification to FAULT for experiment five. First, it became apparent that computer-based training must not only provide test results, but must also provide information on how tests and procedures are performed. When such information was provided in experiment 5, the procedural errors were eliminated and the computer-based training was on a par with, or better than, traditional instruction (Johnson and Rouse, 1982b). Second, it was learned that such global measures as number of problems solved or time to solution are not adequate to discriminate performance differences. Therefore, a method was developed to categorize errors and use error analysis as a finer-grained measure of performance. Such analyses proved to be valuable for this and other such evaluations.

B. Training Ship Engineers

In 1980, the FAULT approach to diagnostic training was adopted by a firm that specializes in training engineering officers of large cargo ships and supertankers. Much of this training had been delivered by classroom lecture and operation of a full-scope ship control room simulator.

FAULT was used to provide subsystem troubleshooting practice to complement use of the full-scope simulator. The most significant effect of this application was that the instructors took responsibility for development and maintenance of the FAULT data bases. This application has continued to prosper, independent of ancillary support from the original developers of FAULT. Further, the parent company of the marine training organization has adopted FAULT for training on aircraft systems. Finally, and perhaps most indicative of the successful acceptance of FAULT, this training organization is now participating in a Navy-sponsored R&D effort aimed at developing and evaluating more sophisticated versions of FAULT (Su, 1985).

IV. SOCBI FOR MILITARY TRAINING

The majority of the early experimentation and evaluation with SOCBI was done within a college-level technical training environment for aviation mechanics.

The first commercial training application was for marine supertanker engineers, most of whom had engineering degrees. The college students and marine engineers were willing to accept the level of abstraction presented via the FAULT displays shown in Figure 2. However, when the research moved to an Army electronics training environment, it became obvious that Army students were not initially willing to accept or adapt to the same level of abstraction used by these other groups. Army students required a training device that had a degree of physical similarity to the real equipment.

The Signal School at Fort Gordon provides interservice training for personnel who operate and maintain telephone switchboards and related communications networks. The course of instruction within which SOCBI was applied lasts 39 weeks, including 3 weeks on the SB 3614 switchboard. The SB 3614 switchboard is a portable communications device used for tactical environments. This course was selected as the focus for the project due to the fact that the SB 3614 represented state-of-the-art digital technology and because the annual student throughput was quite high. Further, nearly 50 percent of the course was centered on diagnostic-related job tasks.

A. Instructor Involvement

The Fort Gordon project presented a potential difficulty that is familiar to many CBI developers. This problem was the instructors' initial attitude toward CBI. The decision to implement SOCBI was made at a reasonably high level of the school administration. However, the development, implementation, and evaluation had to be accomplished at the instructor level. The instructors included civilian and military personnel with an initial outspoken skepticism toward the value of SOCBI. The computers were perceived as a threat to instructor positions and as a replacement for real-equipment systems, already in short supply.

Quality development of SOCBI requires a thorough knowledge of the subject matter, the capabilities of the computer, and the manner in which students learn about the equipment and associated diagnostic job activities. The subject matter experts must be motivated to take an active role in simulation design, data base development, course implementation, and ongoing evaluation and revision. The result of SOCBI development without frequent user formative evaluation (i.e., feedback throughout development) is SOCBI that dies once the research team leaves.

In an effort to win the support of subject matter experts at the Signal School, the development team involved the instructors in all stages of development. The instructors helped to write the simulation specifications and helped design the implementation and evaluation plan. The fact that one member of the research team became quite knowledgeable of the real-equipment hardware further established the instructor–research team working association. The end result was that the instructors as well as the researchers shared in the pride of authorship.

B. Front-End Analysis

The importance of thorough front-end analysis in the evaluation and/or design of training systems cannot be overemphasized. Therefore, the initial phase of this work involved the compilation of all available information related to the task of operating and maintaining the SB 3614 switchboard. It was necessary to become familiar with the basic theoretical as well equipment-specific training for the communications personnel. Knowledge of the aptitudes and expected abilities of the students was also important to assure the appropriate delivery of specialized training.

Familiarization with the switchboard equipment and training course was a two-step process. First the program of instruction and hour-by-hour course outlines were reviewed. Second, and of significant importance, one of the research team members attended the SB 3614 course. The course attendance had several purposes. As a result of course participation the research team developed extensive in-house content specialty and developed a rapport with the course instructors and administrators. The in-house content specialty and positive working relationship with the subject matter experts were critical to the success of the courseware development as previously mentioned. Course attendance served yet another purpose in allowing a review of training from the students' perspective. This perspective was significant in designing the delivery of the courseware.

The front-end analysis therefore permitted the researchers to become familiar with the job, the job training, the instructors, and the student. Beyond this, the analysis permitted the team to identify those portions of the training that could be appropriately supplemented with a mix of computer-based diagnostic training techniques.

C. Simulation Specifications

While the delivery of both specific and general diagnostic training was the overall goal, it was apparent that a sound simulation must permit the student to interact with the computer in much the same manner as with real equipment. For the SB 3614, this requirement dictated that the simulation permit the student to accomplish the following: verify proper installation; operate the equipment, including test functions and call processing; and receive appropriate information when a failure has occurred. The student had to acquire the necessary skills and knowledge to recognize and diagnose a malfunction, identify the subsystem to which a faulty component belongs, and, finally, select the appropriate failed printed circuit card in the subsystem. In addition to these very specific goals, a more general goal was that the simulation should deliver general training that transfers to a variety of other types of real equipment.

D. Equipment and Courseware

The courseware was written in Pascal on a 64K Apple II+. The Apple II systems were configured with dual disk drives and clocks and used 15-in. monochromatic CRT displays. The hardware system was an Army a priori selection. Therefore, from the beginning, the development proceeded in cognizance of the computing hardware constraints. Thus, the graphics, data storage, and processing speed had predefined limits.

Figures 4 and 5 show examples of the real equipment and CRT displays

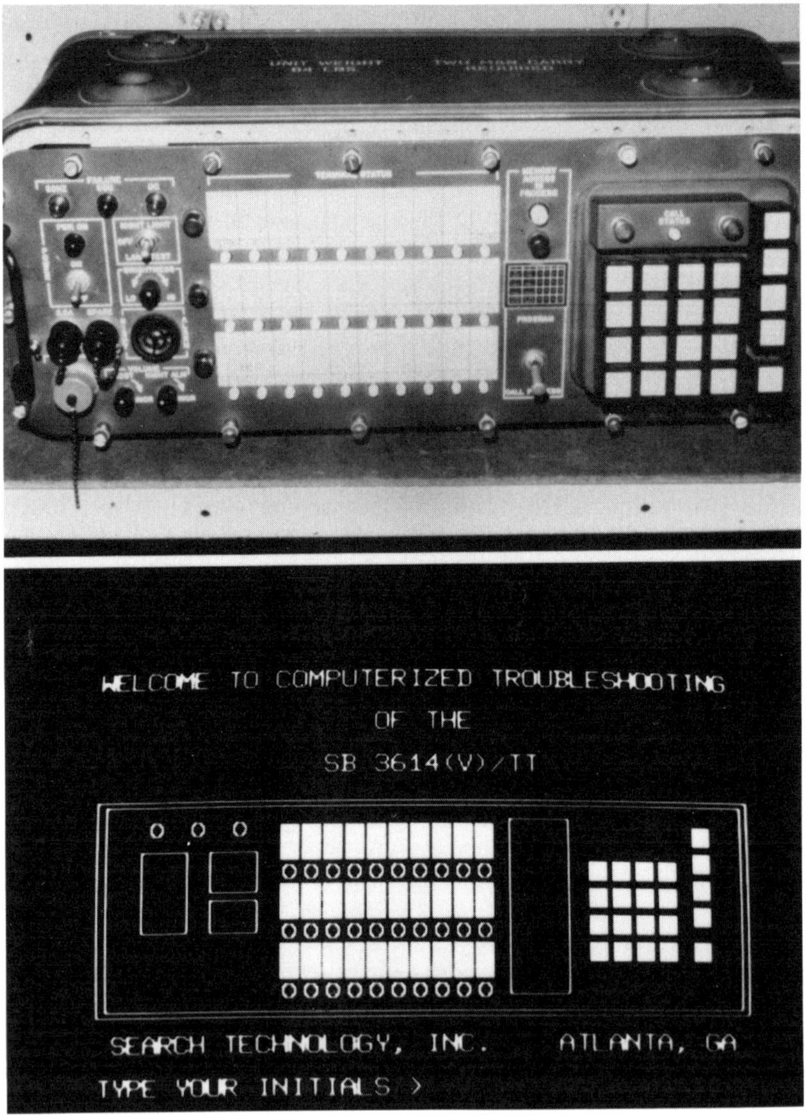

Figure 4. Real and simulated switchboard (SB 3614).

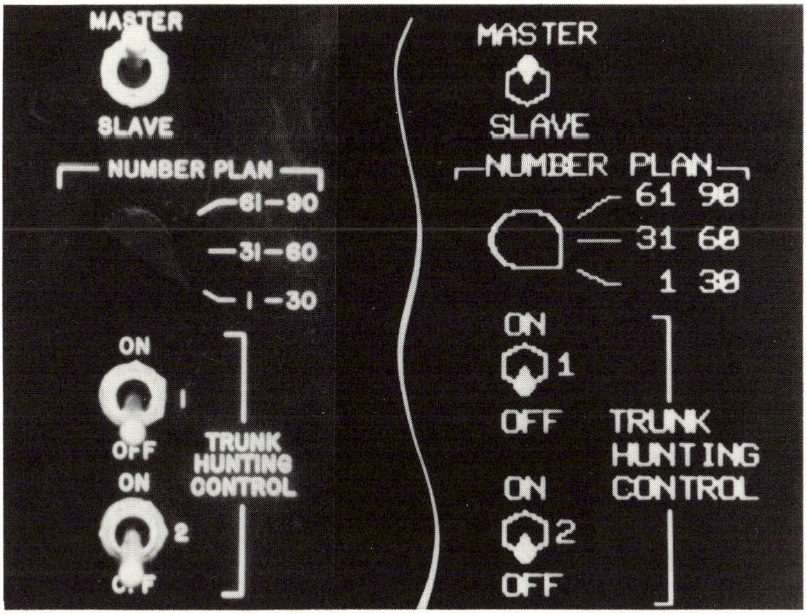

Figure 5. Real and simulated switch panel.

representing that equipment. The simulation mixed these displays with alphanumeric information, including FAULT. While the students initially used the graphically displayed information, it soon became apparent to them that FAULT provided the most information and feedback and therefore eventually received considerable use.

The courseware simulated 19 failure situations and one no-failure condition. The higher-fidelity portion of the simulation allowed students to verify that equipment has been properly installed, to use the test functions, and to process calls. The lower-fidelity portion permitted students to view the system on a conceptual level, using seven FAULT subsystem diagrams to supplement the computer display. It also allowed access to nearly 60 component definitions, as well as information on the easiest way to test each component. A more detailed description of the actual courseware can be found in Johnson and Fath (1985).

E. Evaluation

The simulation-oriented computer-based courseware was integrated into the training environment and evaluated via a pilot test. The control group consisted of 35 Army students, while the experimental group had 30 students. Instructors were the same for both groups. A total of 5 additional hours of instruction was received by both groups beyond that received in previous 3614 classes. The control group had 3 hours of real-equipment practice supplemented by 2 hours of work on paper-and-pencil exercises. The experimental group had a $1\frac{1}{2}$ hour introduction to the courseware and $3\frac{1}{2}$ hours of computer-based problem-solving practice.

Posttraining data collected for both groups related to performance on a live-equipment troubleshooting test. It involved 133 live-system problems solved by 63 students in a 12-hour period. Performance measures were solution success and time to solution for each problem. In addition, a troubleshooting performance checklist was used to record the step-by-step actions of 39 students (20 control and 19 experimental) involved in solution of one particular problem. These checklists were completed by four experts who rated each student action and then assigned a composite rating to the solution as a whole. An overall performance rating was determined using the action-by-action ratings in a manner similar to that described by Johnson (1980). Finally, the experimental group was given a posttraining opinionnaire.

The most appropriate measure of pretraining performance on the SB 3614 was students' scores on a midterm written exam. This exam was administered after the first six days of SB 3614 instruction. The exam was designed to measure the student's overall knowledge of switchboard operation, basic construction, and various subsystem functions. Analysis of the exam scores showed that there was no significant difference between the two groups with regard to the students' basic knowledge and understanding of the switchboard.

The ratings of time and problem solution success were considered to be "global" in the sense that they provided an overview of performance. The time-to-solution comparisons were based on all 133 solutions to three problems. There was no significant difference in the solution times or the solution success between the two groups.

As mentioned in Section II, experience with evaluation of live-system troubleshooting has indicated that global measures such as time and successful problem solution are not, in themselves, the best measures from which to make comparisons between groups (Johnson and Rouse, 1982a). Therefore, a group of four experts participated in a fine-grained analysis of 39 subjects' actions.

While the inter-rater correlation was low, the analysis of the action-by-action checklists showed that the experimental group was always rated as high as the control group. When there were significant differences, they always favored the group trained with the computer-based simulation.

An opinionnaire was given to the students shortly after the live-system troubleshooting test. The instrument was designed to elicit opinion regarding the quality of the simulation as a training device for troubleshooting. Eighty-one percent of the users felt that the computer simulation helped them in real-switchboard troubleshooting. Ninety-six percent felt that the same simulation technique should be used in other courses. One hundred percent reported that they were "satisfied" or "very satisfied" with the simulation as a classroom learning aid.

F. Summary

In summary, the results at Fort Gordon showed that SOCBI was a reasonable substitution for a certain amount of real-equipment training and practice. Students trained with SOCBI had at least the same diagnostic performance, and in some cases, better performance than those trained with real equipment. In a posttraining opinionnaire, the SOCBI users reported a very high level of satisfaction with SOCBI. The instructors also gave the SOCBI high marks. The simulation was used experimentally through 1983, after which time it was officially adopted into the program of instruction, where it continues to be used today.

The evaluation at Fort Gordon provided additional insights that affected subsequent SOCBI developments. First, the extensive use of the simulation by Army students suggested that SOCBI must contain at least a minimal level of face validity (i.e., physical resemblance to the real equipment). This resemblance is necessary to provide the user with a very important initial impression that the SOCBI indeed embodies the characteristics of the real equipment. Once the user begins interaction with the simulation, using familiar cues from real equipment, he/she is more readily willing and able to accept various levels of abstraction (or lower physical fidelity). Ultimately, the SOCBI user will gravitate toward portions of the training that provide advice and feedback over portions that merely "look" like the real equipment. The SOCBI use at Fort Gordon reinforced our strong premise that troubleshooting is predominately a cognitive task, thus minimizing the need for high physical fidelity in diagnostic training simulation. The training personnel there helped coin the phrase that SOCBI provides "heads-on" rather than "hands-on" practice.

A second finding at Fort Gordon centered on the manner in which students learned how to use the simulation. While hard-copy student operating instructions were provided, very few students chose to use them. Instead, students went directly to the computer and "learned by doing." While the on-line instructions were minimal, they seemed to be adequate. As a result, all subsequent SOCBI developments have embodied extensive on-line operating instructions and placed minimal reliance on the use of hard-copy documentation. Such on-line instructions are a must for SOCBI.

V. TRAINING FOR THE NUCLEAR INDUSTRY

The most recent stage of the evolutionary process was the development of a computer simulation for diagnostic training for the nuclear industry. The work, sponsored by the Electric Power Research Institute (EPRI), was divided into three phases. First, a study of the present status of diagnostic training in the nuclear industry was conducted using a combination of on-site interviews and a mail survey. Second, SOCBI was conceptualized and developed working with personnel from a large electric utility. In the third phase, the SOCBI was integrated and evaluated in a training course for nonlicensed operators and maintenance technicians. The entire research effort is fully documented in Volumes 1 and 2 of EPRI Report NP-3829 (Johnson et al., 1985; Maddox et al., 1986).

A. Phase 1—Status of Nuclear Diagnostic Training

An extensive survey (Johnson and Maddox, 1983, 1986; Johnson, Maddox, Rouse, and Kiel, 1984) of trainers from nuclear power plants showed that there was an expressed need for improved diagnostic-related training for both operations and maintenance personnel. In addition to establishing the fact that there was a need for improved diagnostic training, the survey identified the skills and knowledge that trainers considered to be important for competent diagnostic performance. The most important were the following: knowledge of interaction among plant systems, knowledge of specific system operation, and knowledge of general troubleshooting procedures.

B. Phase 2—Development of SOCBI

1. Selecting the Developmental System

With the phase 1 consensus that there was considerable room for improvement in the area of diagnostic training, the second phase set forth to identify a technical system that would serve as the focus of specific diagnostic training research and development. The specific system had to be problematic for the industry as reported by Nuclear Regulatory Commission Licensee Event Reports (LERs). Further, the system had to be general enough that the finished product would be of broad interest to the EPRI utility membership. With these constraints, the emergency diesel generator (DG) was selected as the developmental system.

2. Development Philosophy

The project proceeded with the philosophy that useful computer-based courseware must be developed by competent educational technologists in close harmony with technical instructors and equipment operation and engineering personnel. The team included the author and his colleagues, with backgrounds in human factors, technical training, systems engineering, and computer science, and utility personnel with expertise in nuclear training, nuclear engineering, and mechanical engineering.

3. Design

The simulation development followed a systems approach. The team defined the DG-related diagnostic tasks of the maintenance mechanics and unlicensed operators. Both the utility and the Institute for Nuclear Power Operations (INPO) had performed extensive job and task analyses for the reactor operator and therefore had data bases in which DG-related skills and knowledge were specified. These analyses were also applicable to unlicensed operators and maintenance technicians and therefore used to expedite and enhance the front-end analysis.

In addition to identifying the requisite knowledge and skills for DG diagnosis, attention was given to the identification of the information sources used for DG diagnosis. This attention to information sources focused on the following questions:

- What are the primary avenues for operational and diagnostic information?
- How and where is the information accessed?
- How specific must the information be?
- Who must have access to the information?

The answers to the above questions were derived via discussions with operations, maintenance, and training personnel from the utility's three nuclear stations. The engine controls and instrumentation for the three nuclear stations were also photographed with the intention of identifying the important generic information that must be delivered by the simulation.

When a training simulation is properly designed, it should be adaptable to the knowledge and skills of the user. Therefore, the design criteria for this project specified that the courseware should be useful for operators as well as mechanical maintenance technicians. The simulation was to present mechanics and operators with an opportunity to gather DG information and to make diagnostic-related decisions similar to those decisions made on operational equipment.

In addition to providing the student with an opportunity to practice troubleshooting, the simulation had to provide intelligent feedback whenever the user performed a significant action. Diagnostic error-identification algorithms provide two types of feedback: (1) negative feedback when the user performs an illogical procedure; (2) positive feedback when the problem is solved. Additionally, the user is given performance scores relative to the average for that problem across all previous students. This thorough and intelligent feedback separates SOCBI from the more traditional modes of computer-based instruction.

A final requirement in the design criteria specified that the user be able to learn quickly to use the simulation. Time spent learning the simulation is time lost from learning troubleshooting. As noted in Section IV, previous work revealed a tendency for students to avoid reading the user's manual. Therefore, information was available on-line and easily accessible as needed and requested. Also, the simulation was menu-driven and mouse-operated to eliminate the need to learn commands.

The success at Fort Gordon using computer-generated graphics to deliver the student to logic-based diagnostic training prompted further enhancement for the nuclear training project. The displays are generated on a high-resolution color graphics terminal with a DEC LSI 11/73 processor. The human-computer interface is via a mouse or the keyboard, at the user's discretion. Recently, the

Figure 6. Diesel generator simulation.

program was transported to an IBM PC format (Johnson, Wiederholt, and Maddox, 1986) and made available to all member utilities of EPRI. The demonstration system is shown in Figure 6.

The simulation specifications required that the user be able to access the same information that would be available in the nuclear plant. With this capacity, it should be possible for the student to experience the majority of the cognitive processing that would be experienced during real-equipment troubleshooting.

The main menu of diagnostic information sources is shown in Figure 7. These sources were derived with extensive interaction between the author (and his colleagues) and the training and engineering personnel at the nuclear plant. From this main menu the user has the opportunity to check controls and instrumentation on both the main and the local instrument panel, review system schematics, obtain information from operators, replace parts, or use FAULT. The student chooses the course of diagnostic information by using a mouse.

Figure 8 shows one of the displays that the student can access via the second window in Figure 7. The Main Local Panel display is a generic panel designed via an analysis of similar displays in three nuclear power plants. Special consideration was given to using sound human factors principles in the generic display

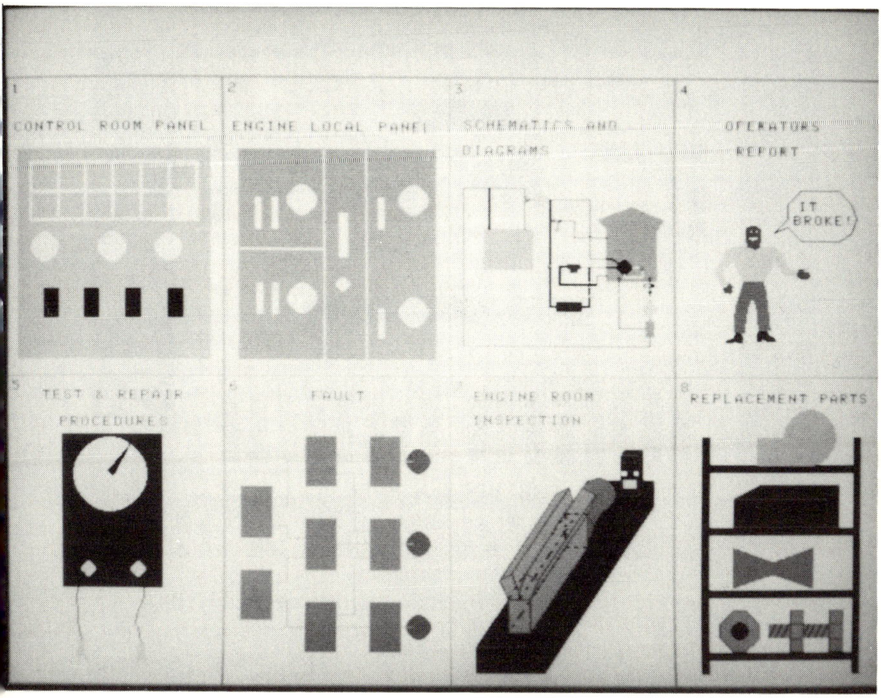

Figure 7. Main menu for diesel generator simulation.

design. In particular, the type of display element(s) (e.g., bar graphs, gauges, digital indicators) used for each parameter was dictated by the way the information from that element was to be used by troubleshooters. These displays are applicable to other equipment merely by changing numbers associated with setpoints and operating ranges.

4. Formative Evaluation

Development of the DG courseware was an iterative process which involved a close working relationship between the software developers and utility personnel. This process contributed to the formative evaluation (i.e., evaluation during formation) of the courseware. The formative evaluation served to assure that the courseware was consistent with the design specifications and provided initial assessments of the "human factors" of the courseware.

The last phase of formative evaluation involved a somewhat formal demonstration at the utility's Technical Training Center. Eight senior and junior instructors used the courseware for a total of 8.7 hours, during which 50 diagnostic problems were completed.

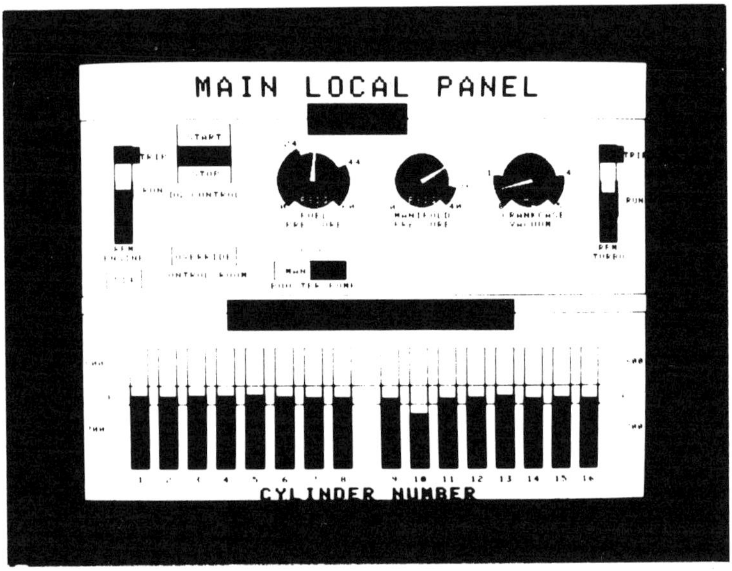

Figure 8. Local panel for diesel generator simulation.

The objectives of this demonstration included determination of the following:

- Extent to which trainees could understand and interact with the courseware
- Adequacy of both off-line and on-line embedded instructions
- Extent to which off-line human assistance is needed during initial use
- Correctness of technical information presented
- Extent to which diagnostic scenarios challenge trainees
- Overall impression of courseware upon initial use

In addition to pursuing these objectives, the instructors were polled relative to their opinion about how the DG courseware could best be integrated with other elements of the utility's DG training. After this initial evaluation, the simulation was prepared for the formal implementation and evaluation.

C. Phase 3—Summative Evaluation in the Nuclear Training Environment

The summative evaluation of SOCBI in the nuclear training environment was the most extensive transfer of training study ever undertaken in the nuclear power industry (Maddox, Johnson, and Frey, 1986; Maddox and Johnson, 1986).

1. Experimental Method

A total of 39 students participated in the final evaluation. Of these, 31 were nonlicensed operators, called nuclear equipment operators (NEOs), who were

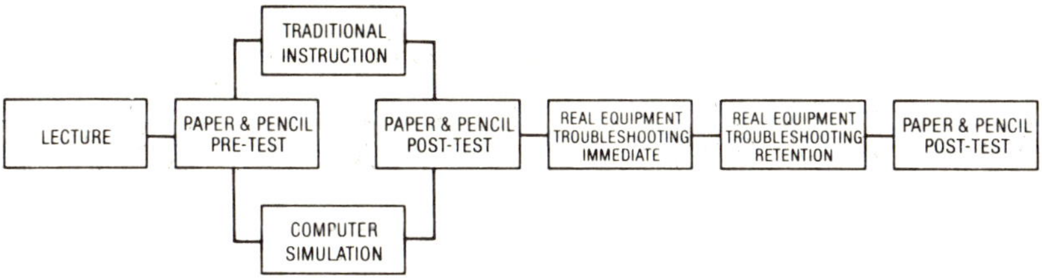

Figure 9. Experimental design for DG evaluation.

participating in "requalification" training. The remaining 8 students were maintenance technicians. The flow of the experimental and control treatments and subsequent data collection is shown in Figure 9.

The utility developed a two-day course segment addressing generalized troubleshooting techniques, emergency DG subsystems, previous problems with DGs, etc. The students were randomly assigned to either the experimental (SOCBI) or control (traditional) group. The first day of training was the same for both groups and was given to the entire class at the same time and in the same classroom. The course material for the first day was related to the overall function of the DG system and to generalized troubleshooting concepts and techniques. After the first day of class, a pretest was administered. The pretest consisted of 30 multiple-choice questions related to DG operation and generalized troubleshooting.

On the second day of the course, the SOCBI group solved simulated problems using the courseware. The problem-solving session lasted approximately 4 hours. For the traditional group, the second day of the course consisted of a very detailed walkthrough of the real DG room with an instructor. During this walkthrough, the instructor described each major DG subsystem and its relationship to other subsystems. The instructor also pointed out where DG problems had occurred in the past and what symptoms one would expect to see if specific components or subsystems failed. The traditional group also received a detailed lecture presentation of the troubleshooting approach embodied in FAULT. This lecture was an attempt to expose the control group to the same concepts as the SOCBI group.

At the end of the second day, both groups came together again to take the written posttest. The posttest consisted of 50 multiple-choice questions, 30 of which appeared on the pretest and an additional 20 related to specific DG and troubleshooting facts taught in the second day of the course.

For the initial transfer-of-training part of the study, five DG problems were selected by a consensus among the utility instructors, operations and maintenance engineers, and the research team. Example failures were the following: open breaker on water heater, failure of main fuel pump, and stuck thermostatic valve on oil cooler. Students were taken individually into the actual DG room at the nuclear plant where an experimenter explained the initial conditions for a postulated failure. The student walked through the DG room and asked for specific information or told the experimenter what tests should be done, etc. The experimenter supplied requested gauge readings, system status, test results, and other information requested by the student while keeping a detailed log of student actions. The students were given a maximum of 20 minutes to locate each of the five postulated failures.

One hundred and forty days after the transfer trials, the students received the posttest again. Also, each student went through the transfer trials with five failures, two of which were identical to failures seen in the original transfer trials. The same experimenter conducted both sets of transfer trials.

Three sets of dependent measures were collected. One set was related to students' use of the SOCBI courseware and consisted of detailed accounts of display access, keystrokes, and times associated with each. Data related to knowledge transfer during training were the pretest and posttest scores.

The final and most important data related to students' performance on the transfer trials. Summary measures, such as average time to solution, number of actions, and number of timeouts were analyzed. Additionally, each student action was rated by the experimenter and two experts, two licensed operators and two senior maintenance instructors. Each action was numerically weighted according to its rating (+2 for most appropriate, +1 for appropriate, . . . , and −2 for most inappropriate) and a summary "performance index" (PI) was calculated for each student on each problem. This rating scheme was a refinement on the methods used with aviation mechanics and Army communications students.

2. *Results*

For the paper-and-pencil posttest score, there was no difference between those trained with SOCBI vs. those receiving traditional instruction. However, for *changes* in score, the traditional training had a significant positive effect which was attributed to specific information offered during the FAULT lecture.

A total of 20 weeks elapsed before the retention effectiveness data were collected. The 50-question posttest was readministered to all students. While there was an overall drop in all scores, there was no differential effect on drop in score due to either student background or training type.

Students' performance on the transfer task (five simulated problems in the real DG room) was analyzed using the following dependent measures:

- Solution time
- Number of actions
- Performance index (PI)
- Experimenter's performance rating (PR)

- Number and types of errors
- "Real-world" time associated with error actions

There were no statistically significant differences in any dependent performance measures attributable *solely* to the difference in training method. While the type of training had no effect on the average student performance index (PI), a more detailed analysis of errors revealed a significant interaction effect in these data. The maintenance students trained with traditional instruction committed serious errors at a rate five times as great as any other group. More specifically, it was found that NEOs performed with low error rates, having had either SOCBI or traditional instruction. For maintenance technicians, however, SOCBI was much more effective in reducing serious troubleshooting errors than traditional instruction. This result has practical consequences in terms of the potential damage such errors can cause in real troubleshooting situations.

For the retention study, the data-gathering and scoring procedure was the same as it was for the problem-solving sessions conducted after initial training. As in the initial sessions, five problems were presented in random order to each student and their diagnostic actions were written down by the experimenter. These actions were subsequently graded by the experimenter and two other subject matter experts from the utility.

In the retention transfer trials, students who received the SOCBI course segment solved the simulated DG problems significantly faster than trainees who received traditional instruction. The SOCBI trainees solved these problems 23 percent faster, on the average, than students who received traditional instruction.

This result is very important, since in the initial transfer task measures, taken immediately after training, there was no difference in solution time attributable to the type of training students received. And yet, 20 weeks later this effect was very pronounced. The simplest explanation for this effect is that the SOCBI training allowed students to retain the troubleshooting expertise acquired during training (and perhaps acquire more expertise from their daily work routine) than did the traditional training. A second interpretation is that the SOCBI-trained group was able to learn more during the first set of plant troubleshooting problems as a result of their training and therefore performed better during the retention evaluation 140 days later. Other measures of performance showed no significant differences during the retention study.

Considering how the simulation was accessed, it is useful to recall that students who received the SOCBI training saw five practice problems and then as many additional problems as time permitted. As noted earlier, each access path was called a "door" and there were a total of eight doors through which students could gather diagnostic information. Maintenance students spent nearly twice as much time per door than did NEOs. Since maintenance technicians were less familiar with the DG than NEOs, it appears that they required more time to collect and process the information presented in each door.

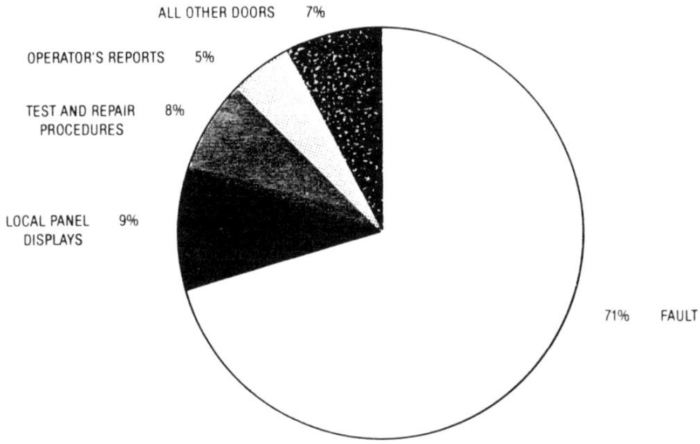

Figure 10. Use of DG simulation.

Students spent more time using the FAULT door than all other doors *combined*. Figure 10 shows that students spent 71 percent of their time in FAULT for each problem. In fact, four of the eight doors (FAULT, Local Panel, Test and Repair Procedures, and Operator's Report) accounted for a total of 93 percent of the average time per problem on the SOCBI courseware.

3. *Conclusions*

Based on this three-year research project, several conclusions can be made concerning diagnostic training in general and the use of SOCBI for nuclear training. Prominent among these findings are the following:

- Diagnostic training benefited nuclear plant personnel with maintenance and operations diagnostic responsibility.
- SOCBI was as good as or better than traditional instruction for diagnostic training.
- SOCBI provided diagnostic practice not available through traditional methods.
- SOCBI should be developed with significant participation from technical instructors who are subject matter experts.
- SOCBI users favored simulation options that provided extensive feedback and intelligent advice over options with less information and more color graphics.

VI. CONCLUSIONS: LESSONS LEARNED AND A LOOK TO THE FUTURE

This chapter has described the first decade of an ongoing research and development effort on computer-based training for troubleshooting. The work began with the attempt to understand human problem solving through basic research, described in Section II, and then continued into a variety of instructional domains, described in Sections III, IV, and V. Each phase of the research provided another piece to a complex puzzle that remains incomplete. However, the research has left a trail of training systems that remain operational today. This final section will reiterate the reasons why this research resulted in not only scientific findings, but also practical operational concepts and systems that continue to flourish.

A. Understanding the Troubleshooting Task

The work with TASK and early work with FAULT demonstrated that logical problem solving can be taught with computer-based methods by permitting the student to practice diagnostic tasks in a variety of contextual domains. The computer-based troubleshooting practice had to challenge the user with the same cognitive task as real-equipment troubleshooting. The simulation had to permit the troubleshooter to gather information, formulate diagnostic hypotheses, and test hypotheses using the same test equipment data available in the real operational environment.

The early research reinforced the assertion that competent technical problem solvers combine logic and knowledge to solve problems. Competent troubleshooting requires a logical approach to information gathering and processing as well as an understanding of each component's operation and the interaction of components in an integrated system. Further, competence is most often a function of experience. Since logic, knowledge, and experience seemed to be the key to diagnostic competence, the computer-based diagnostic training was designed to provide training and reinforcement in these areas.

The computer-based training provided the student with the opportunity to learn about a technical system and to practice troubleshooting in an environment of reinforcement and feedback. The research showed that novice trainees initially need higher fidelity than do more experienced technicians. Experienced troubleshooters accepted higher levels of abstraction (i.e., lower physical fidelity) and seemed to be able to more readily transfer from one diagnostic environment to another.

While the experimental evaluations usually compared traditional instruction with computer-based instruction, the work in marine engineering training and at the Army Signal School suggested that the best training results may be achieved with a proper mix of high-fidelity trainers, including real equipment, and lower-fidelity computer-based trainers. Most importantly, the training had to provide high cognitive fidelity.

B. Design and Development Team

This chapter described four applications of SOCBI. They were aviation in a university environment, marine technical systems, army electronics, and nuclear safety systems. Of these four applications, three continue to use SOCBI in a format similar to the way it was developed to address research questions. One of the four is no longer in use. The original researchers (i.e., Johnson, Rouse, and Hunt) are not present at any of the four locations.

What are the circumstances that prompted three of the four original users to continue while one did not? The answer is straightforward and makes a strong statement regarding the design and development team. The three groups who continue to use the system are those where the technical instructors were actively involved in the conceptualization, design, and data base development. Where the instructors were not involved in the development of the SOCBI, the simulation is no longer being used. Of the many findings in this 10-year research effort, this conclusion is among the most important. The technical instructors must have a key role on the development team. In a CBI development effort where the researchers conceptualize, design, and develop the training without the participation of the technical instructors, use of the resulting simulation is likely to diminish once the research team leaves the training site.

C. Implementation and Evaluation

The experienced gained over the last 10 years in the area of implementation and evaluation can be readily condensed. It is critically important that an organization commit adequate resources to properly implement and evaluate the CBI once it is developed. Too often, funds are committed to an initial development effort but sufficient ongoing support is not provided. Another common occurrence is that the CBI looks good to the managers, instructors, and students; however, there is no attempt to determine whether the CBI is valuable.

D. A Look to the Future

The computer and display hardware that was used at the start of this research, in 1976, is antiquated by today's standards. Likewise, the capabilities of today's hardware will see substantial improvements into the next 10 years. As much as the hardware capabilities have improved, so have user attitudes toward and expectations of computer-based instruction.

Instructional software must continue to make advancements in the area of ease of use and quality of feedback. Designers must continue to develop instructional programs that have more of the attributes of human instructors. Such "intelligent tutors" must have an understanding of the technical systems, an ability to recognize typical learning difficulties, an ability to show the learner where there are apparent problems, and an ability to explain concepts to the learner. It therefore appears that the greatest improvements in computer-based training over the next decade must be in the area of artificial intelligence, that is, developing programs striving for the characteristics of expert human instructors.

While the new technology permits us to have great expectations for the future, particularly for intelligent computer-based instruction, we should also look to the centuries of successful apprenticeship training. The master provides instruction to the apprentice and incrementally permits the apprentice to perform greater amounts of the job task. The master of today must be the intelligent job aid. Intelligent portable job-aiding computers have the potential to redefine on-the-job training, particularly for diagnostic tasks. As this author speculates on the opportunity to report this evolution of diagnostic training 10 years hence, I expect to be writing about the integration of training and aiding for improving human effectiveness in diagnostic tasks. The last decade of research and development has clearly set the stage for such integration.

ACKNOWLEDGMENTS

The research described here was sponsored by the Electric Power Research Institute, the U.S. Army Research Institute for the Behavioral and Social Sciences, Marine Safety International, Inc., and the National Aeronautics and Space Administration. The author gratefully acknowledges the training personnel from the U.S. Army Signal School at Fort Gordon and from the technical training center of Duke Power Company, Charlotte, NC. Additional thanks to colleagues at Search Technology including Janet Fath, Ruston Hunt, Michael Maddox, William Rouse, Bradley Wiederholt, Anita Smith, and Kim Goulette.

NOTE

1. Of course, there is a long history of research on fault diagnosis or troubleshooting, as attested to by the review efforts of Standlee and his colleagues 30 years ago (1956). More recent reviews can be found in Johnson et al. (1980) and Morris and Rouse (1985).

REFERENCES

Alessi, S.M., and Trollip, S.R. 1985. *Computer-Based Instruction-Methods and Development.* Englewood Cliffs, NJ: Prentice-Hall, Inc.

Hunt, R.M. 1979. *A Study of Transfer of Problem Solving Skills from Context-Free to Context-Specific FAULT Diagnosis Tasks.* Urbana: University of Illinois, Coordinated Science Laboratory, Report T-82.

Hunt, R.M., and Rouse, W.B. 1984. A fuzzy rule-based model of human problem solving. *IEEE Transactions on Systems, Man, and Cybernetics SMC-14:*112–120.

Hunt, R.M., and Rouse, W.B. 1981. Problem solving skills of maintenance trainees in diagnosing faults in simulated powerplants. *Human Factors 23*(3):317–328.

Johnson, W.B. 1980. Computer simulations for fault diagnosis training: an empirical study of learning transfer from simulation to live system performance. Doctoral dissertation, University of Illinois. [*Dissertation Abstracts International* 1981, *41*(11):4625-A (University Microfilms No. 8108555).

Johnson, W.B. 1984. Simulation-oriented computer based instruction for fault diagnosis. In *Simulation and Training Technology for Increased Military Systems Effectiveness.* Warrenton, VA: Society for Applied Learning Technology.

Johnson, W.B., and Fath, J.L. 1983. Design and evaluation of mixed-fidelity courseware for maintenance training. *Proceedings of the 27th Annual Meeting of the Human Factors Society.* Norfolk, VA: Human Factors Society, pp. 1017–1022.

Johnson, W.B., and Fath, J.L. 1985. *Implementation of a Mixed-Fidelity Approach to Maintenance*

Training. Alexandria, VA: U.S. Army Research Institute for the Behavioral and Social Sciences, Report TR-661.

Johnson, W.B., and Maddox, M.E. 1983. Status of diagnostic training in the nuclear utility industry. *Proceedings of the 27th Annual Meeting of the Human Factors Society*. Norfolk, VA: Human Factors Society, pp. 151–155.

Johnson, W.B., and Maddox, M.E. 1986. Development, implementation, and evaluation of computer-based simulation for operations and maintenance troubleshooting training in nuclear power stations. In *Advances in Human Factors in Nuclear Power Systems*. LaGrange, IL: American Nuclear Society, pp. 428–439.

Johnson, W.B., Maddox, M.E., and Kiel, G.C. 1984. Simulation-oriented computer-based instruction for training nuclear plant personnel. *Proceedings of the 28th Annual Meeting of the Human Factors Society*. San Antonio, TX: Human Factors Society, pp. 1008–1012.

Johnson, W.B., Maddox, M.E., Rouse, W.B., and Kiel, G.C. 1985. *Diagnostic Training for Nuclear Plant Personnel,* Vol. 1: *Courseware Development*. Palo Alto, CA: Electric Power Research Institute, Report EPRI NP-3829.

Johnson, W.B., Rouse, S.H., and Rouse, W.B. 1980. *An Annotated Selective Bibliography on Human Performance in FAULT Diagnosis Tasks*. Alexandria, VA: U.S. Army Research Institute for the Behavioral and Social Sciences, Report TR-435. (ERIC Document Reproduction Service No. ED 192 736.)

Johnson, W.B., and Rouse, W.B. 1982a. Analysis and classification of human errors in troubleshooting live aircraft power plants. *IEEE Transactions on Systems, Man, and Cybernetics* *12*(3):389–393.

Johnson, W.B., and Rouse, W.B. 1982b. Training maintenance technicians for troubleshooting: two experiments with computer simulations. *Human Factors 24*(3):271–276.

Johnson, W.B., Wiederholt, B.J., and Maddox, M.E. 1986. *Diagnostic Training Demonstration: Instructor and Student Manuals and Sample Diskettes*. Palo Alto, CA: Electric Power Research Institute, Report EPRI NP-4493P.

Maddox, M.E., and Johnson, W.B. 1986. Can you see it? Can you understand it? Does it work? An evaluation plan for computer-based instruction. In *Advances in Human Factors in Nuclear Power Systems*. LaGrange, IL: American Nuclear Society, pp. 380–389.

Maddox, M.E., Johnson, W.B., and Frey, P.R. 1986. *Diagnostic Training for Nuclear Power Plant Personnel,* Vol. 2: *Implementation and Evaluation*. Palo Alto, CA: Electric Power Research Institute, Report EPRI NP-3829.

Morris, N.M., and Rouse, W.B. 1985. Review and evaluation of empirical research on troubleshooting. *Human Factors 27*(5):503–530.

Richardson, J.J., Maxion, R.A., Polson, P.G., and De Jong, K.A. 1985. *Aritifical Intelligence in Maintenance: Synthesis of Technical Issues*. Brooks Air Force Base, TX: Air Force Human Resources Laboratory, Report AFHRL-TR-85-7.

Rouse, W.B. 1979a. Problem solving performance of maintenance trainees in a fault diagnosis task. *Human Factors 21*(2):195–203.

Rouse, W.B. 1979b. Problem solving performance of first semester maintenance trainees in two fault diagnosis tasks. *Human Factors 21*(5):611–618.

Rouse, W.B. 1982. A mixed-fidelity approach to technical training. *Journal of Educational Technology Systems 11*(2):103–115.

Rouse, W.B. 1984. *Computer-Generated Display System Guidelines:* Vol. 2: *Developing An Evaluation Plan*. Palo Alto, CA: Electric Power Research Institute, Report EPRI NP-3701.

Rouse, W.B. 1983. Models of human problem solving: detection, diagnosis, and compensation for systems failures. *Automatica 19*(6):613–625.

Rouse, W.B., and Hunt, R.M. 1984. Human problem solving in fault diagnosis tasks. In W.B Rouse (Ed.), *Advances in Man-Made Systems Research,* Vol. 1. Greenwich, CT: JAI Press

Standlee, L.S., Popham, W.J., and Fattu, N.A. 1956. *A Review of Troubleshooting Research.* Bloomington: Indiana University, Research Report No. 3.

Su, Yuan-Liang David. 1985. *Modeling Fault Diagnosis Performance on a Marine Powerplant Simulator.* Atlanta: Center for Man–Machine Systems Research, Georgia Institute of Technology, Report No. 85-3.

GRAPHIC INTERFACES FOR SIMULATION

James D. Hollan, Edwin L. Hutchins, Timothy P. McCandless, Mark Rosenstein, and Louis Weitzman

ABSTRACT

The dynamic graphic display of the state of complex systems has immense potential value for mediating the development of richer understandings of process as well as for providing more effective mechanisms of interaction. For the past few years we have been involved in the development of a set of software tools to assist in the construction of interfaces to simulations and real-time systems. These tools have been used extensively in the development of an interactive inspectable simulation-based instructional system, STEAMER (Hollan, Hutchins, and Weitzman, 1984). Underlying our efforts are three interrelated research activities: formulating a theory of interface design, understanding the effectiveness of interactive graphic representational systems, and implementing systems based on these developing theoretical notions. The dialectic between these activities has been very valuable for us

Advances in Man-Machine Systems Research, vol. 3, pages 129–163

ISBN: 0-89232-659-X

as cognitive scientists and as system builders. In this chapter, we survey some of the presuppositions of our approach to interface design, describe the tools we have built to assist in the construction of graphic interfaces, and discuss the conclusions we have drawn from our experiences in graphic interface design.

I. INTRODUCTION

To illustrate the type of graphic interface with which we are concerned, we begin by briefly describing STEAMER, a system which employs an interactive graphical interface to a steam propulsion plant simulation. STEAMER is a research project involved with evaluating the potential training applications of techniques from the new disciplines of artificial intelligence (AI) and cognitive science. While the project addresses a host of research issues ranging from how people understand complex dynamic systems to how AI software and hardware advances might be applied to training, it is focused around the construction of a computer-based system to assist in propulsion engineering instruction. The goal of the project is not only to build a training system with tutorial and explanation facilities but also to construct a set of software tools to assist in the implementation of simulation-based training systems and graphical interfaces.

STEAMER currently consists of a graphic interface to a mathematical model of a steam propulsion plant. The interface allows a user to select from a library of propulsion plant views and interact with a selected view to change the state of the underlying simulation model. The evolution of plant states can be observed by graphic changes in the view on a color display. Views depict aspects of the propulsion system at various levels of detail. They vary from collections of gauges and indicators typically found in a real plant to schematic diagrams specifically designed to depict models similar to those experts seem to employ in reasoning about the operation of the propulsion plant. The potential for increased instructional effectiveness derives from representations with the ability to show global views of systems which are physically dispersed in the actual plant, and thus difficult to see as a total system, to those with the ability to show simplified versions of systems designed to be easier to understand or to provide better models for reasoning about the plant, to look "inside" systems or components and see flows or other internal characteristics, and to make available indicators which depict aspects of the operation of a system not normally available but which are useful in developing an understanding of a system.

Figures 1–4 are black-and-white renditions of views a user of STEAMER would see on a color graphics screen. State information is depicted by color, by animation, and by analog, digital, and textual changes. For example, operational status of a pump or valve is indicated by color (red for off; green for on); flow rates in pipes are dynamically shown by use of animation techniques; dials and graphs reflect plant parameters. The iconic representation serves both to provide state information and as a mechanism for changing state of an underlying simula-

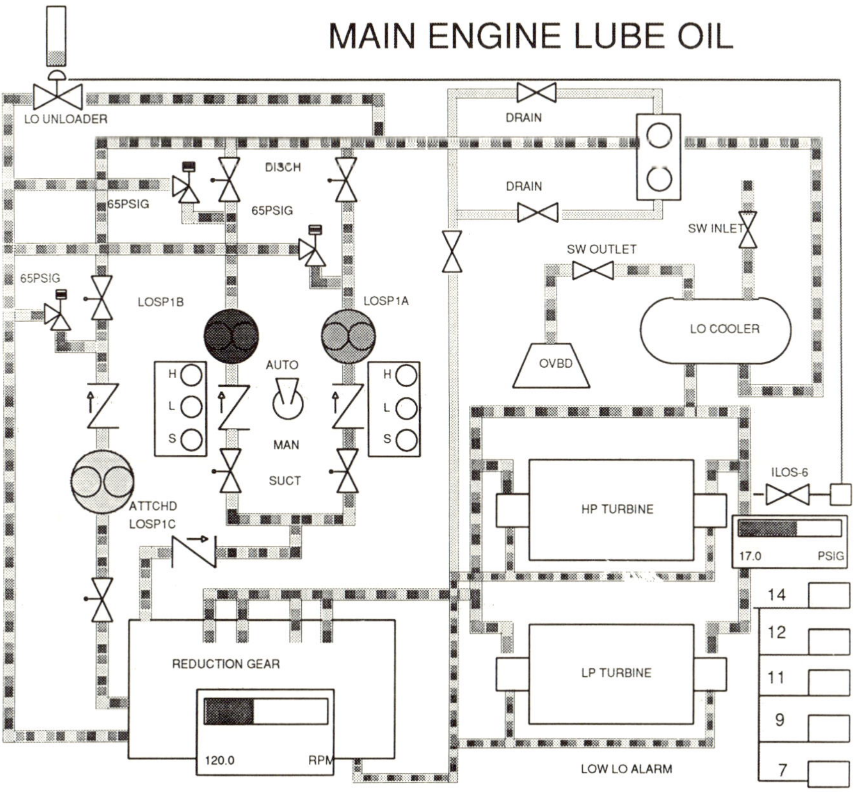

Figure 1. Main engine lube oil. The lube oil system is distributed throughout the propulsion plant. This view provides a global view of its operation. Like most figures in this chapter, it is a black-and-white rendition of a view normally presented on a color screen. Dithering techniques have been used to map from colors to stipple patterns. The boxes shown within pipes are used on the color screen to provide animation of flows.

tion. By pointing to a component with a mouse-controlled cursor, a user can change its state by clicking on it. For example, clicking on a pump will toggle its state. Similarly, one can vary the level of a tank, change the value of a dial, or position a throttle. Of course, the nature of the underlying simulation and the goals of the interface designer determine which variables and thus which components can be manipulated by a user. The important point here is that the interface functions as a two-way communication device: depicting and allowing changes of state.

A high-level view of the main engine lube oil system is depicted in Figure 1. One can watch the state of the lube oil system and its responses to changes made

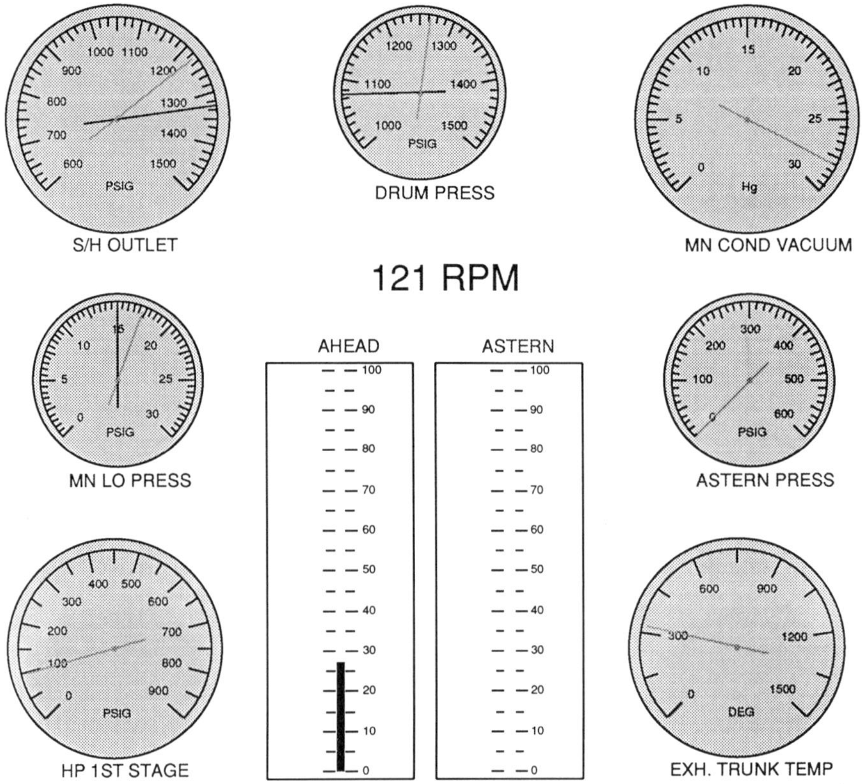

Figure 2. Throttle board. This view shows the following major system parameters: superheater outlet pressure, mainsteam drum pressure, main condensate vacuum, astern pressure, exhaust trunk temperature, high-pressure first-stage turbine pressure, and main lube oil pressure; also, it allows for control and monitoring of the ahead and astern throttles.

to the propulsion system. The flows in the system are dynamically depicted by animation within pipes and the connectivity of the system is shown. The states of the lube oil service pumps (LOSP1A-C) are indicated by color changes. In Figure 1 the attached pump LOSP1C is operating, LOSP1B is off, and LOSP1A is operating at low speed (colors in the figure are represented by differing gray shades). A series of pressure sensors is shown at the lower right of the view along with digital and analog representations of lube oil pressure. These sensors monitor pressure at the bearing most distant from the pumps and will automatically control the state of the service pumps if pressure drops below specified levels. The valve at the top left is the lube oil unloader valve and the column above it indicates how far it is open. As pressure in the system rises above a set

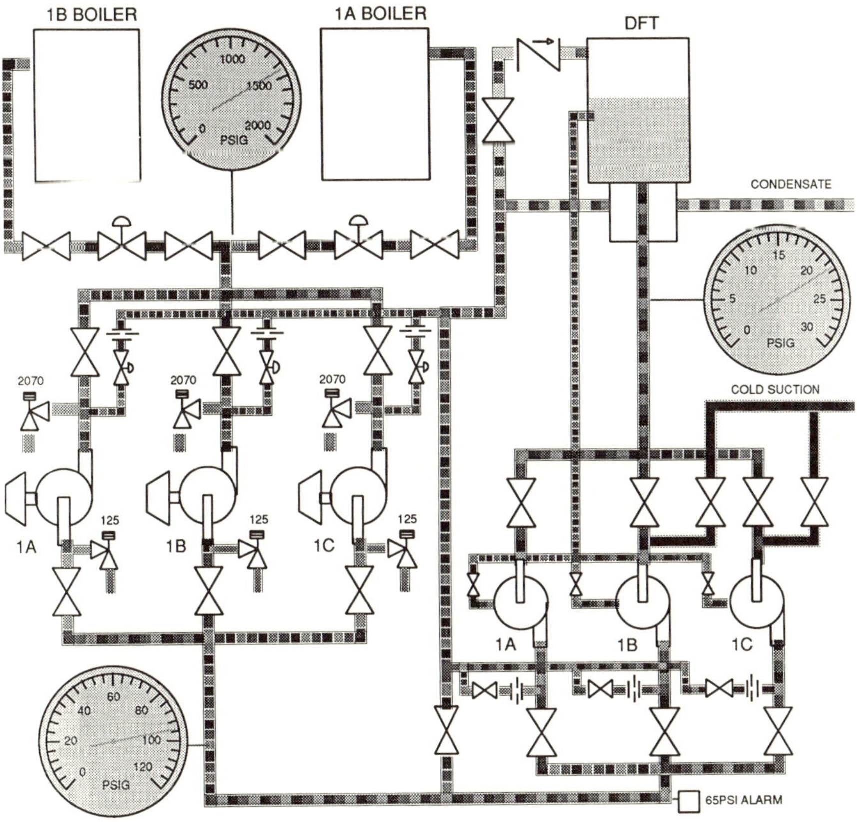

Figure 3. Main feed system. This shows the topology and major components of the feedwater system.

threshold, the unloader valve opens and unloads lube oil back into the sump, preventing overpressurization. There are two controller boxes and a switch next to lube oil service pumps 1A and 1B. By touching the switch the system can be put in manual mode, and the controller boxes for the pumps can be touched and operated to change pump speed to high (H), low (L), or stop (S). As the controller is operated, the associated pump will change state and the ramifications of that change will be continuously reflected in other portions of the view.

Figure 2 shows a throttle-board view that allows the user to control the ahead and astern throttles and monitor a number of important system parameters. Figures 3 and 4 depict portions of the feed system. Figure 3 shows the states of the two boilers (1A and 1B), the level of the deaerating feed tank (DFT—a water storage tank), the states of the six pumps involved in the system, and the large

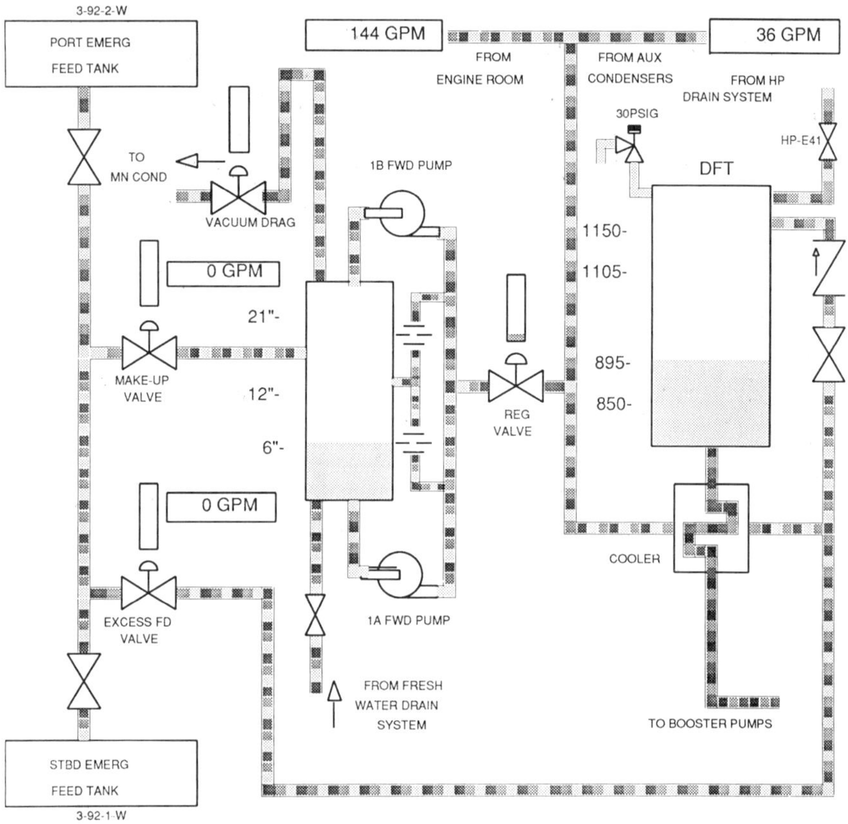

Figure 4. Makeup and excess feed. This is another portion of the feed system designed to show control relationships between tank levels and states of associated valves and pumps.

number of valves used to control and direct the feed water to the boilers, as well as key system parameters.

The same type of graphic interface can be provided for a real-time simulation. For example, Figure 5 contains a view we designed to monitor the dynamic state of a computer on our local area network. It shows the number of users running various programs, and continuously graphs interrupts, system calls, CPU idle time, characters in and out, users on the system, and processes waiting to be run.

These examples are intended to illustrate the type of interactive graphic interfaces with which we are concerned. The key notion is that the interface serves as a communication device to allow a user to see and manipulate the state of an underlying simulation or real-time system. In the STEAMER application we have been concerned with using this form of interface in training. It should be clear that it is just as applicable for monitoring or controlling a system. Our

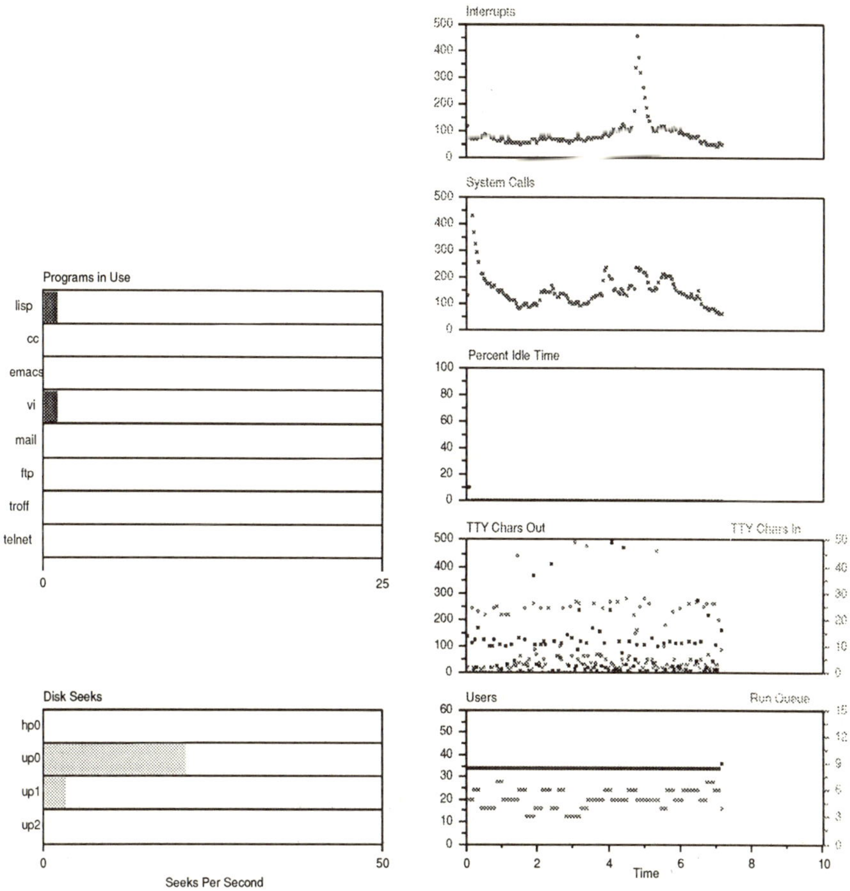

Figure 5. Unix display. This is a view designed primarily to illustrate the connection of a graphic view to a real-time system. It provides a summary of the state of an operating system running on one of the computers on our local area network. (Unix is a trademark of Bell Laboratories.)

primary concerns here are to discuss why we think this form of interactive graphic interface is powerful and describe a set of tools we have implemented to facilitate interface implementation.

II. UNDERLYING PRESUPPOSITIONS

A common way of describing the class of interfaces discussed above is as *direct manipulation* interfaces (Shneiderman, 1982). In our view, most of the treatments of direct manipulation interfaces focus on the wrong level of analysis. The naive notion seems to be that the key properties are bit-mapped displays and pointing devices. One of our presuppositions is that interfaces are representational systems designed for communication, and just as in the analysis of any other representational system it is essential to understand the cognitive task that the system is attempting to support. It is an egregious error to suppose that one can discuss interface design outside of the cognitive contexts of the task domain in which an interface is embedded. The *directness* associated with an interface comes from how directly the interface supports the user's cognitive task. Graphic displays and pointing devices are media of support, but do not in themselves guarantee any directness. Directness results when the interface language closely matches the way in which a user thinks of a task. Directness is thus not a property of interfaces but involves the relationship between the task a user has in mind and the way in which the task can be accomplished via the interface.

An interface provides a language for the user to communicate with a system and for the system to communicate with a user. A key notion for us is the relationship between the meaning of an expression in the interface language and what the user wants to say. We have termed this relationship *semantic distance* (Hutchins, Hollan, and Norman, 1986). The extent to which an interface language allows one to say what one wants to say without circumlocutions is the measure of its semantic directness. Another aspect of directness is the relationship between the meanings of expressions in the interface language and their physical forms. We call this *articulatory distance* (Hutchins, Hollan, and Norman, 1986). Nonarbitrary relationships between meaning and the way it is physically expressed provide this form of directness. A nonarbitrary graphical relationship between icons and what they depict is an example of articulatory directness. Similarly, on the input side of an interface language, interfaces that allow one to make statements about position by pointing provide examples of articulatory directness.

Thus, one of the important underlying presuppositions of our approach is to view an interface as a representational language and to place primary importance on the cognitive task that a user is attempting to accomplish. Another fundamental presupposition of our approach is the view that graphic forms of representation provide powerful ways of bringing abstract things into the realm of the perceptually knowable. We think there are important cognitive properties which make graphic representations effective. These properties derive from the types of processing activities people are especially good at: detecting patterns, constructing mental models or simulations of the world that support causal reasoning, and manipulating the world by actions on it or on representations of it.

Rumelhart, Smolensky, McClelland, and Hinton (1986) have argued that one of the effective problem-solving strategies people employ involves the creation of artifacts, physical representations that can be manipulated to get answers to questions. They suggest that the underlying strategy is to make "problems conform to problems we are good at solving" and argue as follows:

> We are especially good at pattern matching. We seem to be able to quickly "settle" on an interpretation of an input pattern. This is an ability central to perceiving, remembering, and comprehending. Our ability to pattern match is probably not something which sets humans apart from other animals, but is probably *the* essential component to most cognitive behavior.
>
> We are good at modelling our world. That is, we are good at anticipating the new state of affairs resulting from our actions or from an event that we might observe. This ability to build up expectations by "internalizing" our experiences is probably crucial to the survival of all organisms in which learning plays a key role.
>
> We are good at manipulating our environment. This is another version of man-the-tool-user, and we believe that this is perhaps the crucial skill which allows us to think logically, do mathematics and science, and in general to build culture. Especially important here is our ability to manipulate the environment so that it comes to represent something. This is what sets human intellectual accomplishments apart from other animals (Rumelhart et al., 1986, pp. 44–45).

If these are indeed the types of activities that people are especially good at, there are clear implications for why graphic interfaces may have important cognitive properties. They provide a physical representational system which permits us to make abstractions perceptually available and thus allow the use of our powerful pattern-matching abilities. They make possible the depiction of models of the world which are similar to the mental models or simulations people seem to use to reason about the world. These models can depict physical state information, causal connections, and are runnable, permitting a user to see the effects that result from changes of state. Finally, graphic interfaces provide the potential of directly manipulable representations of systems. These factors comprise another set of presuppositions underlying our approach to graphic interface design and also provide motivation for our interest in simulation-based systems.

III. TOOLS FOR CONSTRUCTING GRAPHIC INTERFACES

There is a need for software tools to assist in the creation of graphic interfaces. Here we describe a set of tools we have been evolving to facilitate interface design. First we describe a general simulation environment which consists of a *model controller* and a *graphics editor*. This is the core of the system we are developing. It permits one to build interactive graphic interfaces to simulation models or real-time systems. The graphics editor makes available a set of icons, facilities for modifying characteristics of icons (e.g., size, shape, color, and placement), and the ability to associate icons with model variables so that the icons reflect the values of the variables and so that one can interact with the icons to change the values of their associated variables. The model controller allows one to run simulation models, observe state via graphic views constructed with the graphics editor, and interact with the views to change the state of a simulation model.

We also will describe a number of related tools we are developing to support the construction of interfaces. Chief among them is an *icon editor,* which makes possible the construction of new icons without requiring a user to operate at the level of code writing. In addition we discuss a series of *knowledge-base editors* for the specification of domain knowledge. We have implemented a *behavior editor* to explore the incorporation of simulation knowledge into icons and are in the process of developing a *lesson editor* to explore the incorporation of domain knowledge into graphic views so that they can explain themselves, pose problems to students, and monitor their answers. *Designer,* an interactive visual design consultant for users of the graphics editor, makes available graphic design knowledge during the process of constructing and critiquing graphic views. Underlying a number of these knowledge-base editors is a frame-based representational language. We will also briefly describe it and a graphic interface to it.

A. Simulation Environment

The simulation environment we have designed consists of a set of activities to allow users to build, observe, and manipulate views of a simulation model or real-time process. In our work we have used this facility to build interfaces to mathematical simulations, such as the steam propulsion simulation used in STEAMER, real-time systems, and parallel distributed processing (PDP) models learning to recognize patterns in the operation of underlying systems. A system consists of a process and a set of user-defined views connected to that process.

One interacts with a simulation system via its associated views. A *view* is a graphic collection of icons representing a portion of a simulation. We have designed a *model controller* to allow users to manipulate views, observe the effects of the manipulations, and control the underlying simulation model. Using this controller, a user selects two views. One is typically a view used to control global aspects of a system, while the other is used to manipulate and observe subsystems. In STEAMER, for example, the global view contains the throttles for the ship and important status information and the other view can be alternated between any of the approximately 100 views available.

Each simulation environment activity, such as editing or running a simulation model, is supported by a screen configuration which provides a set of functions and menus to help a user accomplish the tasks associated with the activity. Integrated into the current simulation environment are the model control and graphics editor activities as well as a related set of activities discussed below.

B. Model Control

The *model control* activity provides facilities for changing systems, models, and views. It also allows modification of the behavior of a model, such as the rate at which it runs in the case of a simulation or the rate at which an interface is sampled in the case of a real-time system. Figure 6 shows a typical configuration for the model controller. The status line near the bottom of the screen maintains current state information. In this case the STEAMER system has been selected and its associated model is running with the makeup and excess feed view

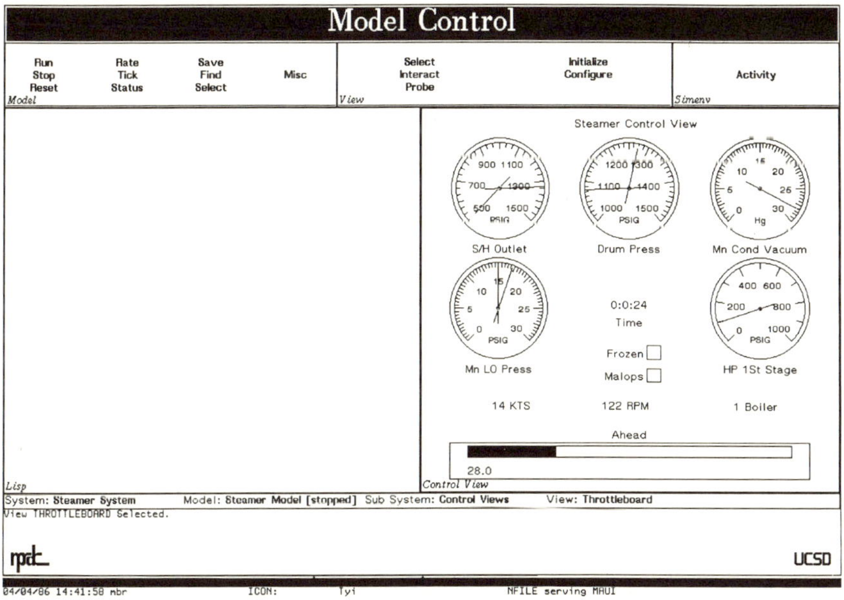

Figure 6. Model controller. The screen configuration of the black-and-white display during operation of a simulation. It provides functions for controlling the running of a simulation, switching between graphic views, and selecting other activities. It is typically used while viewing and interacting with a view on the color screen.

displayed on the color screen. The right half of the middle section of the screen shows a STEAMER control view. At the bottom of that view is the ahead throttle and immediately above that are important data, such as the ship's speed, engine rpm, and the fact that the ship is currently operating with one boiler. Above this information are other indicators of the global state of the propulsion system. Across the top of the screen are menus of operations available for controlling the model and views. The rightmost menu choice allows changing activities.

The status line allows a standard interface to a set of general control functions. The way to view the status line is as a display of attribute value pairs that describe the system at a given time. The attributes explicitly displayed are the current system, model, subsystem, and view. An asterisk is used to indicate a value that has been modified. Whether a model is running or not is shown in square brackets.

C. Graphics Editor

The graphics editor originated from our work on STEAMER (Hollan, Hutchins, and Weitzman, 1984) and the requirement to implement a large number of dynamic views of a propulsion system. We needed a tool which would allow instructors who were knowledgeable about a domain but computer-naive to create graphic interfaces. The resulting graphics editor has been used to create more than 100 views for STEAMER and has been designed to allow its easy extension into other domains. The editor provides a user with a set of icons which can be arrayed on a graphics screen to create a view of an underlying simulation or real-time system. It provides functions commonly available in computer-aided design systems. One can save and restore views from files, mark the elements of a view (individually, by type, within an area, etc.), and edit those marked elements (moving, copying, deleting, changing color, size, etc.). A grid facility is provided to assist in accurately positioning icons within a view.

The multipaned menu interface to the editor activity is shown in Figure 7. The status line in the lower portion of the screen provides control facilities similar to those in the model control activity. In constructing a view, a user chooses icons from the menu of icons and positions them on the color screen. Figure 8 depicts a subset of the available icons. Typically a process of incremental refinement of the view takes place in which icons are moved around, reshaped, colored, and given labels and appropriate units.

An important characteristic of the graphics editor is the way in which it supports the association of icons to underlying variables in a mathematical model

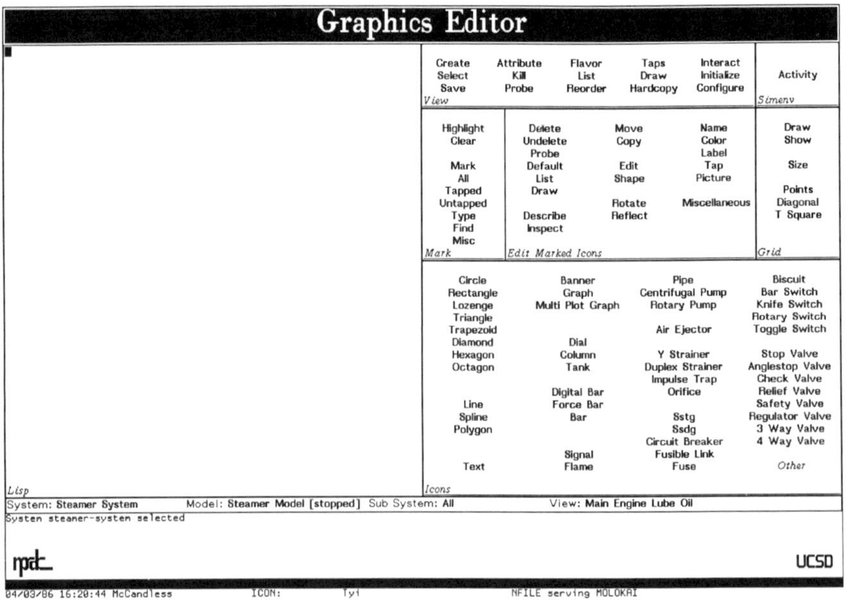

Figure 7. Graphics editor. Screen configuration of the black-and-white display for editing a graphical view.

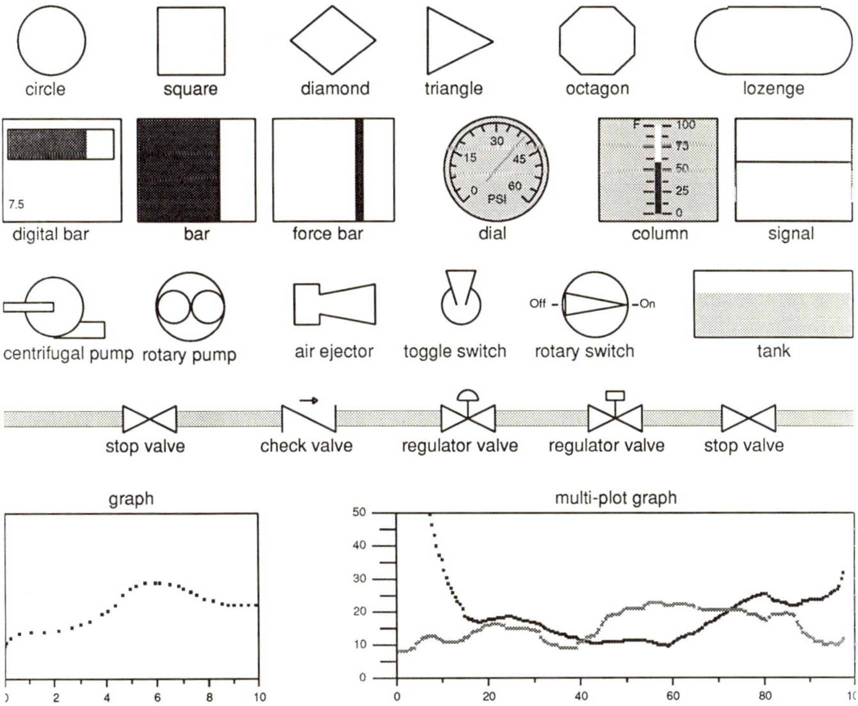

Figure 8. Icon sampler. A subset of the graphical icons available for use with the graphics editor.

or real-time system. We call this process *tapping*. There are two types of taps. A *probe* tap associates a variable with an icon so that the icon reflects the current state of that variable. A *set* tap also associates a variable with an icon but in a way that allows one to change the value of the variable by interacting with the icon. Figure 9 illustrates the tapping mechanism and a simple math model. On the left are the variables in a simple math model and on the right are icons to which they have been tapped. The toggle switch turns this model on and off. Each tick of the clock indicates that one unit of time has passed in the simulation. While the model is running, the valve sets its variable. This means that when one interacts with the column above the valve, it sets the value of its variable, %-Valve-Open, to reflect the valve's degree of openness. The pipe, on the other hand, probes its value, Pipe-Speed. This means that on each tick of the clock, the pipe reflects the value of Pipe-Speed by changes in flow rate. When one interacts to change the state of the valve the effect propagates via the math model and affects pipe flow. Also included in the simulation are the clock and toggle switch themselves. The clock probes its value, Clock-State, while the toggle switch sets its value, Model-Running.

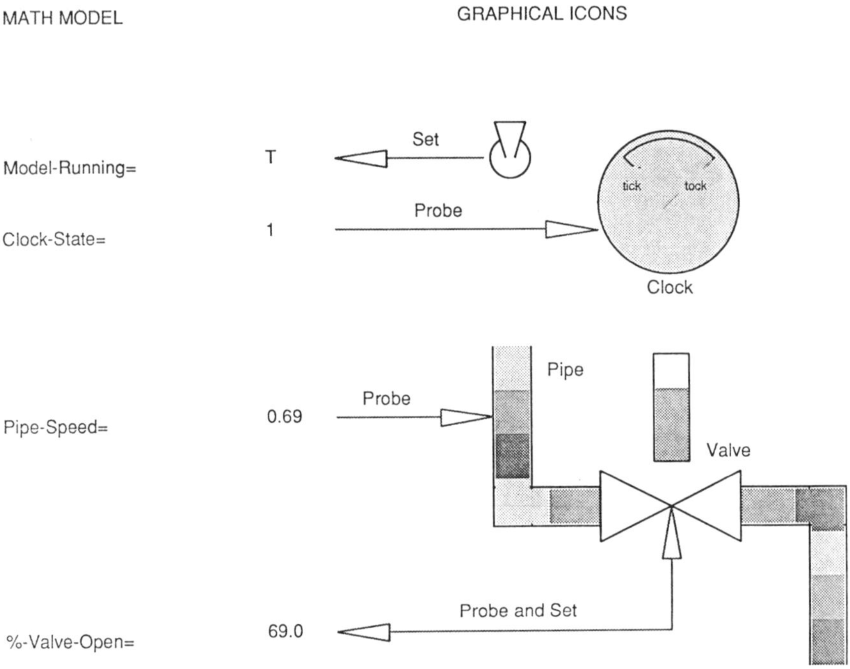


```
;;; Model Augment For TAPPING MECHANISM Diagram
(defun tapping-mechanism-model-augment ()
  (when model-running
    (if (< percent-valve-open 4)
       (setq percent-valve-open 0))
    (setq pipe-speed (// percent-valve-open 100.0))
    (if (= clock-state 0)
       (setq clock-state 1)
       (setq clock-state 0))
    (setq valve-state (if (= 0 pipe-speed) :off :on))))
```

Figure 9. Tapping mechanisms. A depiction of the relationships between variables in a simple mathematical model and a set of icons. *Set taps* allow the value of a variable to be changed by interacting with an icon. *Probe taps* cause icons to reflect the value of associated variables. At the bottom of the figure is a simple model augment used with this view.

To set up the tappings of the pipe, a user of the graphics editor would mark the pipe icon, then click on tap. This generates a pop-up menu (shown in Figure 10) for specifying the tapping parameters. Here the tap probe has been associated with the variable Pipe-Speed. This will cause the pipe to probe that variable when run. Similarly, values of other tapping parameters are provided. The variable %-Valve-Open would be entered as the tap set for the column associated

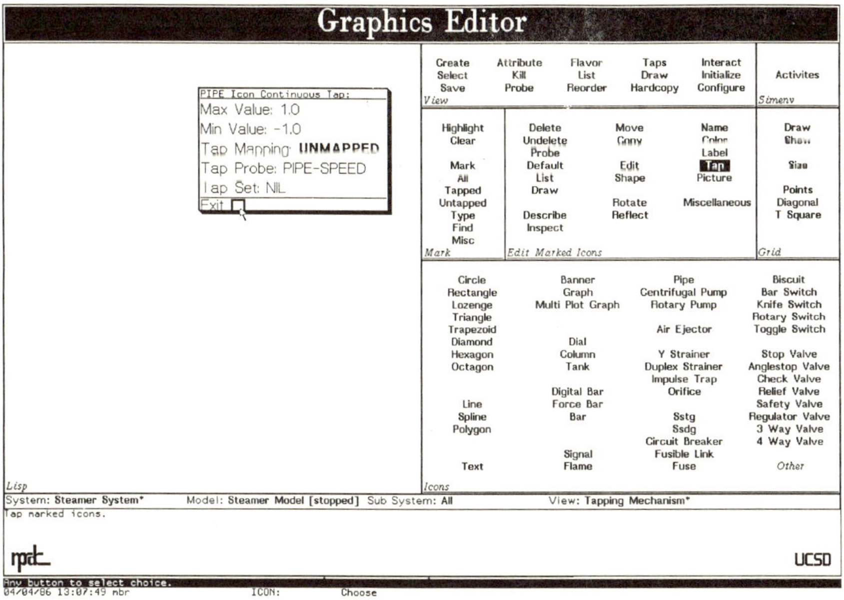

Figure 10. Tapping an icon. Clicking on the tap menu item pops up a window in which probe and set variables can be specified as well as constraints and mappings of their values.

with the valve so that it would be set in the model when a user interacted with it. Notice the tap mapping line for the toggle switch in Figure 11. The mapping mechanism allows the designer to map the icon's value from one scale to another. Selecting *logical* would map the ''on'' state to true and the ''off'' state to nil. Another choice is binary, which maps the ''on'' state to 1 and the ''off'' state to 0. All icons that depict a state or change a state of a math model variable are tapped in this manner.

An associated facility to aid the tapping process is the model augment. There are occasions when a math model insufficiently represents aspects of a simulation that a designer might wish to display. An augment allows one to enhance the simulation model where it is inadequate, to conveniently write more complex tapping code, and to provide stand-alone simulations for views. For example, the STEAMER math model does not repesent flow in pipes. Since we think that flow is important in understanding causality in a steam system, we often add a model augment to a view so the iconic representations of pipes depict flow rates. A model augment also allows the builder to derive values from existing variables. In the simple model augment shown in Figure 9 for the pipe and valve system, we wanted to relate pipe flow to the state of the valve: the more open the valve,

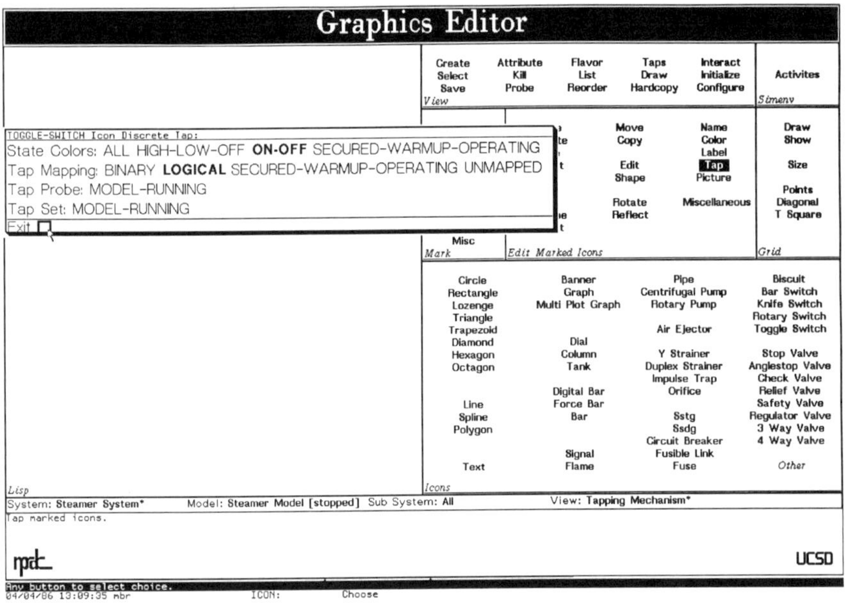

Figure 11. Mapping a tap. The tapping mechanism allows the mapping of state colors and variable types. Here a mapping has been made between a logical variable and the on–off state.

the more flow. The model sets the pipe speed to be the openness of the valve divided by 100.

In designing and implementing the editor we have capitalized on the flexible object-oriented Flavors System of Zetalisp. What is created as a result of the view construction process is a Lisp program which contains a number of dynamic entities capable of responding to messages and of providing graphical support to an interactive instructional system. For example, consider a dial. Many of the properties of the dial actually come from simpler objects which can be used in a variety of icons. Figure 12 shows some of the component pieces of a dial: Rectangular, Continuous, and Tap mixins. For any icon that displays continuous values, we mix in the object called Continuous. This object provides a place to hold the icon's value and the minimum and maximum values the icon can display. The Continuous mixin also provides commands, or messages any instance of a dial icon can understand. Constrain-Value is an example message. It provides a way of constraining a number to be between a minimum and maximum value. Figure 13 shows the complex structure of a dial, including its instance variables and the messages that it can handle. Of course, a user of the graphics editor does not need to think in terms of these implementation details but need only be concerned with critiquing visual characteristics of the dial and tapping it into the appropriate variable.

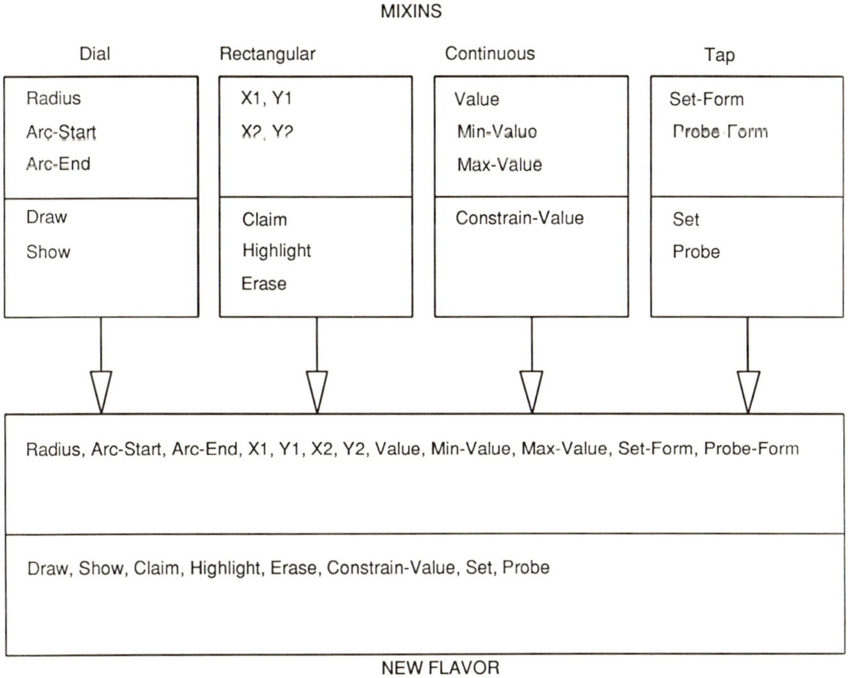

Figure 12. Components of a dial icon. Some of the *mixins* which make up a dial icon. Each mixin contributes instance variables (top half of boxes) and messages (bottom half of boxes).

By having knowledge stored locally in icons, the editor and model controller do not need to know about how the dial does its work. Since the editor does not need to know, neither does the view builder. Icons such as dials or pipes understand other basic messages like Show which, when received, causes the receiving icon to show the value provided in the message. If we send a dial the message to show the value 7, it reflects that value by positioning its needle. On the other hand, if we send a pipe the same message, it shows this value as flow. In neither case did we explicitly need to know how the icon worked, only that it understood the message sent. These messages make possible a very powerful generic interface ability which has been exceedingly useful in the development of the simulation environment.

D. Icon Editor

While the graphics editor allows one to construct views from an existing set of icons, the *icon editor* allows users to construct new icons which can dynamically display and modify the state of a simulation. A user of the icon editor specifies both the icon's appearance and behavior. This specification determines the

DIAL

all instance variables (43)

```
ARC-END (DIAL)
ARC-START (DIAL)
DIAGRAM (BASIC-ICON)
DX (RECTANGULAR-MIXIN)
DY (RECTANGULAR-MIXIN)
FACE-COLOR (GAGE-MIXIN)
FRACTIONAL-CHANGE-TO-SHOW (CONTINUOUS-MIXIN)
FRAME (PICTURE-MIXIN)
INVERSE-MATRIX (RECTANGULAR-MIXIN)
LABEL-COLOR (GAGE-MIXIN)
LABEL-COLOR (RECTANGULAR-MIXIN)
LABEL-FONT (RECTANGULAR-MIXIN)
LABEL-ORIENTATION (RECTANGULAR-MIXIN)
LABEL-POSITION (GAGE-MIXIN)
LABEL-POSITION (RECTANGULAR-MIXIN)
LABEL-STRING (RECTANGULAR-MIXIN)
LOCATIONS (DISPLAY-MIXIN)
MATRIX (RECTANGULAR-MIXIN)
MAX-VALUE (CONTINUOUS-MIXIN)
MIN-VALUE (CONTINUOUS-MIXIN)
NEEDLE-COLOR (DIAL)
OUTLINE-COLOR (RECTANGULAR-MIXIN)
RADIUS (DIAL)
RANGE (CONTINUOUS-MIXIN)
REDLINE-VALUE (DIAL)
TAP-MAP (TAP-MIXIN)
TAP-MAPPING (TAP-MIXIN)
TAP-PROBE (TAP-MIXIN)
TAP-READING (TAP-MIXIN)
TAP-SET (TAP-MIXIN)
TIC-LABELS-COLOR (GAGE-MIXIN)
TIC-LABELS-FONT (GAGE-MIXIN)
TICS (GAGE-MIXIN)
UNITS-COLOR (GAGE-MIXIN)
UNITS-FONT (GAGE-MIXIN)
UNITS-STRING (GAGE-MIXIN)
VALUE (CONTINUOUS-MIXIN)
```

all handlers (70)

```
:ANIMATE primary daemon ICON:DISPLAY-MIXIN
:ASPECT-RATIO primary daemon ICON:SQUARE-ASPECT-RATIO-MIXIN
:BORDER primary daemon ICON:RECTANGULAR-MIXIN
:BOUNDING-RECTANGLE primary daemon ICON:RECTANGULAR-MIXIN
:CHANGE-FLAVOR primary daemon ICON:BASIC-ICON
:CLAIM notype or ICON:RECTANGULAR-MIXIN
:CLAIM-RECTANGLE primary daemon ICON:RECTANGULAR-MIXIN
:COMPUTE-MAPPING primary daemon ICON:TAP-MIXIN
:COMPUTE-PROBE primary daemon ICON:TAP-MIXIN
:COMPUTE-SET primary daemon ICON:TAP-MIXIN
:COMPUTE-TAP primary daemon ICON:TAP-MIXIN
:CONSTRAIN-VALUE primary daemon ICON:CONTINUOUS-MIXIN
:DISPLAY-PICTURE primary daemon ICON:PICTURE-MIXIN
:DRAW combined daemon ICON:DIAL
:DRAW-LABEL primary daemon ICON:RECTANGULAR-MIXIN
:DRAW-NEEDLE primary daemon ICON:DIAL
:DRAW-TIC-LABELS primary daemon ICON:DIAL
:DRAW-TICS primary daemon ICON:DIAL
:DRAW-UNITS primary daemon ICON:DIAL
:DRAW-VALUE primary daemon ICON:DIAL
:EDIT-POSITION primary daemon ICON:RECTANGULAR-MIXIN
:ERASE combined daemon ICON:DIAL
:ERASE-LABEL primary daemon ICON:RECTANGULAR-MIXIN
:EXTENDED-BOUNDING-RECTANGLE primary daemon ICON:RECTANGULAR-MIXIN
:GET-ALL-INSTANCE-VARIABLES primary daemon ICON:BASIC-ICON
:GET-COMPONENT-FLAVORS primary daemon ICON:BASIC-ICON
:HIGHLIGHT combined daemon ICON:DIAL
:HIGHLIGHT-MOMENTARILY primary daemon ICON:DISPLAY-MIXIN
:HIGHLIGHT-POINTS primary daemon ICON:RECTANGULAR-MIXIN
:INIT combined daemon ICON:DIAL
:INVERSE-PROBE primary daemon ICON:TAP-MIXIN
:INVERSE-SHOW primary daemon ICON:CONTINUOUS-MIXIN
:LABEL-POSITION-AUX primary daemon ICON:RECTANGULAR-MIXIN
:LATCH primary daemon ICON:RECTANGULAR-MIXIN
:LEGAL-LABEL-POSITIONS primary daemon ICON:NO-CENTER-LABEL-MIXIN
:LOCATION primary daemon ICON:DISPLAY-MIXIN
:MAKE-PLIST primary daemon ICON:BASIC-ICON
```

```
XC (RECTANGULAR-MIXIN)
XL (RECTANGULAR-MIXIN)
XR (RECTANGULAR-MIXIN)
YB (RECTANGULAR-MIXIN)
YC (RECTANGULAR-MIXIN)
YT (RECTANGULAR-MIXIN)
```

```
:MODIFY-MIX primary daemon ICON:BASIC-ICON
:MOVE-RELATIVE primary daemon ICON:RECTANGULAR-MIXIN
:NORMALIZE primary daemon ICON:RECTANGULAR-MIXIN
:PROBE primary daemon ICON:TAP-MIXIN
:PROBE? primary daemon ICON:TAP-MIXIN
:READING primary daemon ICON:TAP-MIXIN
:PROBE primary daemon ICON:TAP-MIXIN
:PROBE? primary daemon ICON:TAP-MIXIN
:READING primary daemon ICON:TAP-MIXIN
:RECOMPUTE-DERIVED-PARAMETERS combined daemon ICON:DIAL
:REFLECT combined daemon ICON:DIAL
:REMAKE-FORM primary daemon ICON:BASIC-ICON
:SET primary daemon ICON:TAP-MIXIN
:SET-BOUNDING-RECTANGLE primary daemon ICON:RECTANGULAR-MIXIN
:SET-LOCATION primary daemon ICON:DISPLAY-MIXIN
:SET-TAP-MAPPING combined daemon ICON:DIAL
:SET-TAP-PROBE combined daemon ICON:DIAL
:SET-TAP-SET combined daemon ICON:DIAL
:SET? primary daemon ICON:TAP-MIXIN
:SETUP combined daemon ICON:DIAL
:SETUP-COLOR primary daemon ICON:SETUP-MIXIN
:SETUP-COLOR-VALUES primary daemon ICON:DIAL
:SETUP-LABEL combined daemon ICON:DIAL
:SETUP-MISCELLANEOUS primary daemon ICON:SETUP-MIXIN
:SETUP-MISCELLANEOUS-VALUES primary daemon ICON:DIAL
:SETUP-MOVE combined daemon ICON:DIAL
:SETUP-PICTURE primary daemon ICON:PICTURE-MIXIN
:SETUP-POSITION combined daemon ICON:DIAL
:SETUP-ROTATION combined daemon ICON:DIAL
:SETUP-TAP combined daemon ICON:DIAL
:SHOW combined daemon ICON:DIAL
:SHOW-NEW-VALUE? primary daemon ICON:CONTINUOUS-MIXIN
:TAPPED? primary daemon ICON:TAP-MIXIN
:THINGS-TO-SAVE combined append ICON:DIAL
:TO-SHOW primary daemon ICON:DIAL
:ZOOM combined daemon ICON:DIAL
```

Figure 13. Dial icon details. A listing of the 43 instance variables and 70 messages actually contributed to a dial icon by the full set of its mixins.

graphic states for the icon. For example, we have seen the appearance of a dial icon in a number of views. The behavior of the dial icon, like a real gauge, comes from the ability to position its needle. For the icons of the graphics editor these specifications were made by directly writing Lisp code. Experience with the graphics editor has demonstrated that incremental specification and refinement of the appearance of a view is most easily effected through the graphic manipulation of its components, the icons. The icon editor mimics this facility by allowing the specification of an icon's appearance through graphic manipulation of the components of the icon, and specification of its behavior through a critiquing process. This permits the development of a hierarchy of graphic primitives for use in the appearance of icons and a set of graphical behaviors for icons. Thus, the icon editor is a tool for the construction of icons without explicit programming.

A number of the STEAMER icons can be thought of as combinations of existing icons. The digital bar icon in Figure 14 consists of a bar icon, a banner icon, and a rectangle icon. To allow positioning these components, the same graphic techniques available in the graphics editor, such as marking and moving, are also available in the icon editor. The top of Figure 15 shows a flame icon. In a number of views flickering of this icon is used to depict the flicker of an actual flame. To construct the icon with the icon editor, four lines were incrementally added. The icon editor thus makes possible the incremental design of new icons from basic components.

When a user builds a view in the graphics editor the system is writing code. This level of activity is entirely invisible to the user. When constructing a view,

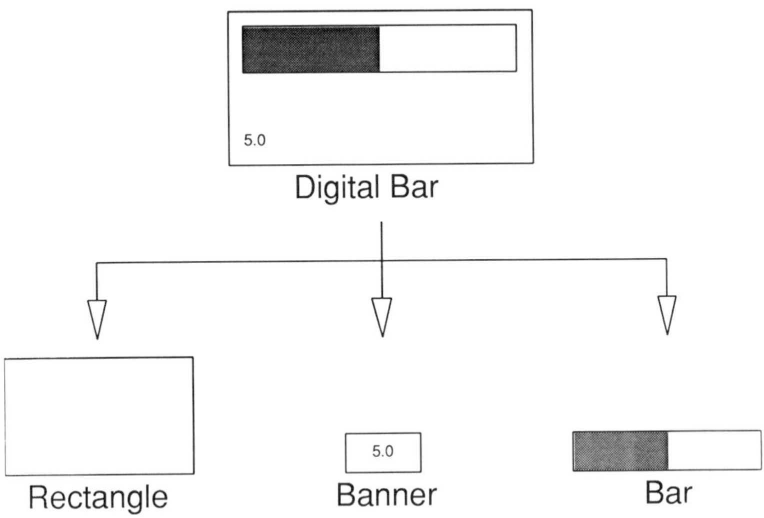

Figure 14. Components of a digital bar.

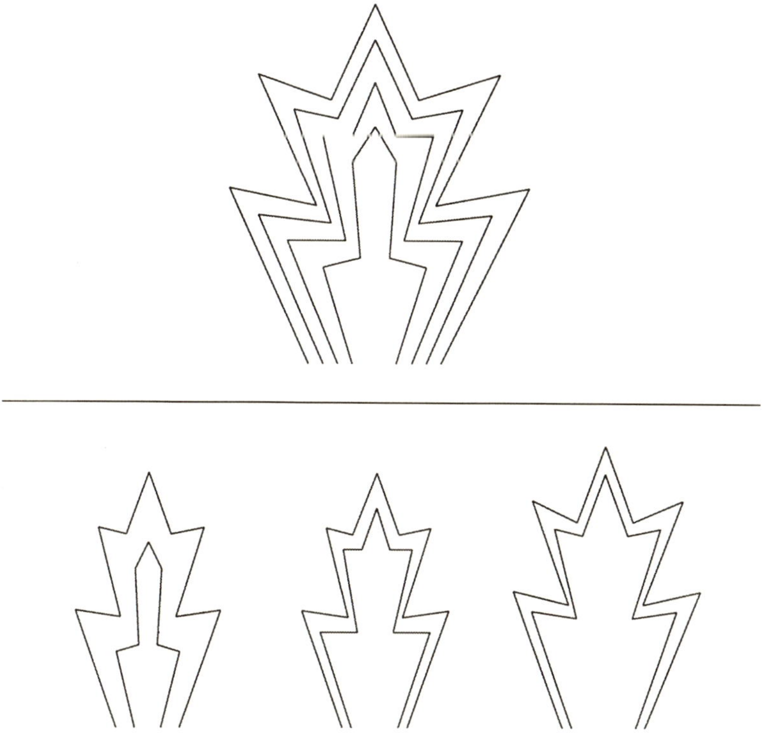

Figure 15. Flame icon. The top portion of the figure shows the complete icon. At the bottom is an animation sequence.

the user has no feeling of coding, of describing the procedure the computer will follow to reconstruct and run the view. Similarly with the icon editor we don't want a person constructing an icon to show that she or he is writing code. Thus, the icon editor must make available an appropriate toolkit of behaviors with semantically direct representation for each.

It is unlikely that there is one general scheme to allow builders of icons access to primitive graphic behaviors. In the digital bar case, the component icons do all the work. The builder specifies the behavior of the digital bar by constraining its tap to be the taps for the bar and banner. In addition, the designer of the digital bar icon must decide which attributes to make available to the user of the icon. In the icon editor a pane of attributes for each component is displayed. The user can constrain two attributes to be identical, in a manner similar to the way taps were constrained, or place an attribute in the color or miscellaneous menus for use in the graphics editor. For the digital bar, the bar's color and the banner's text color would be placed in the color menu. Since bar, rectangle, and banner have labels, the icon builder would not put any of the label attributes of the components in the menu. This prevents the user of the icon from accidentally placing labels in incorrect positions. The icon editor also provides color and miscellaneous buttons, to allow the builder to check the appearance of the these menus as they are built.

The flame icon provides a more interesting case, one in which behavior is synthesized at the level of the new icon. The flickering action of flames is implemented using animation. Rotating the colors of the four lines which comprise a flame creates the flickering effect. A better effect is produced by maintaining a fixed color for one of the lines. At the bottom of Figure 15 are the three possible animation states. When rotation repeats in real time the flame appears to be flickering. Underlying the animation is a complex negotiation between the icon and the graphics device, but the builder needs only to specify the three rotation colors.

Figure 16 shows an experimental configuration for the icon editor in which a flame is being designed. Near the bottom of the figure is the status line showing the current icon (flame). Across the middle of the screen is a pop-up menu an icon user would see in the graphics editor for specifying color. The topmost pane allows a icon builder to examine other graphics editor menus created from the construction of an icon.

We are in the process of implementing additional behaviors for the icon editor. These include allowing an icon to move along a trajectory and to vary the hue of

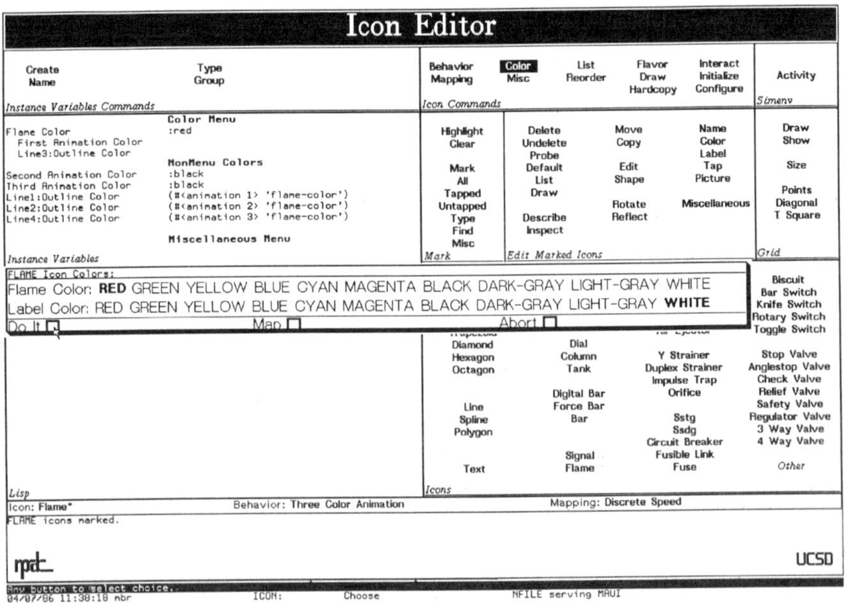

Figure 16. Icon editor. Screen configuration of the icon editor.

its color to reflect the value of a tapped variable. The initial goal for the icon editor is to be able to reimplement the STEAMER icons using the icon editor. We are using this reimplimentation to assist us in identifying a set of primitive components sufficiently rich to permit specification of the diverse set of behaviors exemplified in the STEAMER icons and as a method for exploring interface techniques for making the primitive behaviors available to a user.

E. Knowledge-Base Editors

We are in the process of designing and implementing a set of editors to assist an instructional designer or other simulation interface builder in specifying knowledge about a system. A data base of knowledge is required to integrate information about both the domain and the purposes of particular interfaces. This knowledge can be employed to allow simulation views to be able to describe themselves, run scenarios of simulation activities, pose questions to be answered by interacting with a view or collection of views, and to perform other forms of instructional activities. To facilitate the specification of the data base of knowledge we have designed four additional editors: a knowledge editor with a graphic interface, a lesson editor, a behavior editor, and a graphic design editor.

1. Knowledge Editor and Grapher

We are using a frame-based knowledge representation facility originally designed by Bruce Roberts of Bolt Beranek and Newman, Cambridge, Mass. It essentially builds a class structure on top of Flavors to provide frame-based representational facilities. The underlying language is called MSG, for its flavor-enhancing capability. Here we present an overview of its representational capabilities and our implementation of a graphic interface to it (see Figure 17). The MSG language provides the facility to define classes of objects. Each class defines local attributes that distinguish it from the other classes. It is the instances of these classes that we use to represent the objects in a world being modeled. In the hierarchy of the class structure, a class may have any number of abstractions or superior classes. A given class will inherit all of the attributes of its abstractions. By defining additional attributes at the local level, a class can be made more specific. The inheritance mechanism allows the inheritance of roles and slots. A role is the semantic organization of a set of attributes, while a slot provides an actual placeholder for an attribute. In addition, the system provides a *coreference* facility, the ability to reference the same attribute of a class using multiple descriptions.

When a new class is defined, an instance of a metaclass is created that will hold all pertinent information about the class. This includes how to create a new class instance, where to store these instances, and how to manipulate them. When a new class is defined, a new flavor is also defined. The name of this new flavor is the same as the name of the class being created. When instances of a class are created, an instance of the associated flavor is made. The instance

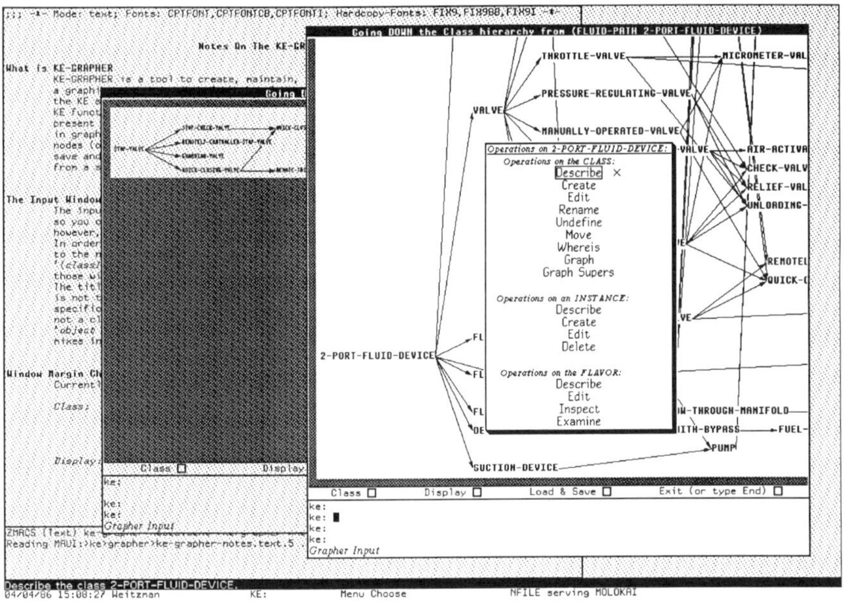

Figure 17. Graphic interface for MSG and Flavors system. A graph of the dependencies between classes. Individual nodes in the graph are mouse sensitive. Clicking on an item gives a menu of actions that can be performed on it.

variables of a class provide the typical *role* and *slot* descriptors of a frame-based representational language. A role consists of a role name and a list of slots which make up the role. A slot provides a name and the potential of specifying restrictions and default values. In addition, the language provides for subroles to further refine roles. Figure 18 shows an example of the MSG representation of a two-port fluid device.

To access the value of a slot of an instance, a path to that slot's value is constructed. For example, the class of fuel-oil-service-pump has an instance called *fosp-alpha* and a slot called *inlet-valve*. To return the fosp-alpha's inlet-valve you would construct the path (*the fosp-alpha inlet-valve*). This would return the object that is fosp-alpha's inlet-valve. If this value happens to be another object one can access a slot of the new object by adding onto the path. If the inlet-valve has a slot called *inlet* you could expand the path to (*the fosp-alpha inlet-valve inlet*), which would return the inlet of the fosp-alpha's inlet-valve.

An important feature of the MSG language is coreference, the ability to reference the same class element by different descriptors. Paths and synonyms play an important role in providing this facility. When a synonym is defined, a coreference is built and used instead of the actual object. The coreference includes the actual object and the multiple paths to access the object. The data

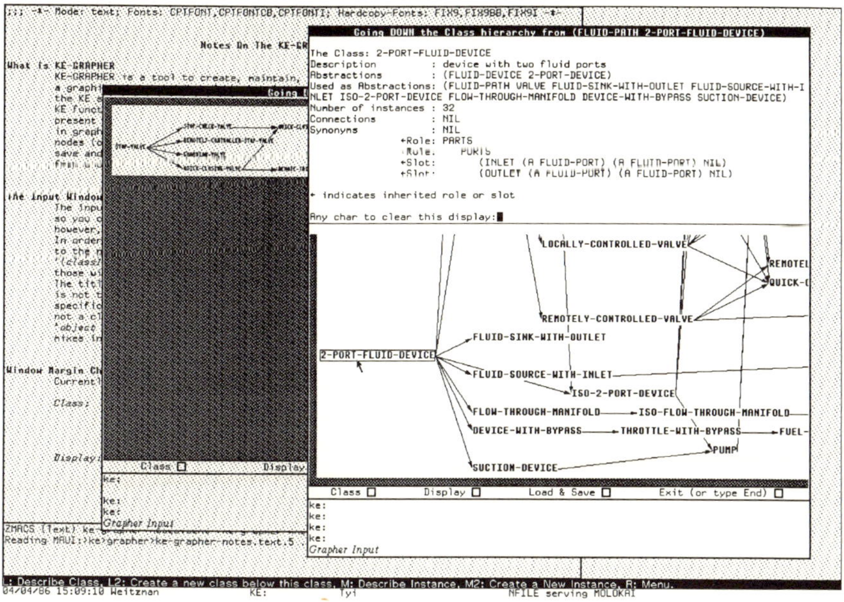

Figure 18. MSG representation of a two-port fluid device. This results from pointing to 2-port-fluid-device and asking for a description of the class.

structure of a synonym is a list of synonym pairs. Each member of a pair defines a path to a slot which is synonymous with the slot found at the end of the path of the other member of the pair. For example, the synonym ((*inlet-valve inlet*) (*suction*)) states that the inlet of the inlet-valve is the same as the suction slot.

KE-GRAPHER is a tool to create, maintain, and inspect MSG objects using a graphic representation of the knowledge class hierarchy. It incorporates a graphing facility to display the class hierarchy with the nodes of the graph becoming mouse-sensitive. Examples are shown in Figures 17 and 18. The window is divided into four parts: the title, the main graphing area which presents the class hierarchy, a margin choice area to select top-level commands, and a keyboard input pane to change the objects to be graphed. The starting nodes of graphs are called *roots*. After a user types in an expression that will evaluate to the name of a class or a list of classes, the window will reset the roots and regraph the window. If the item evaluated is not the name of a class but the name of a flavor, then that flavor and those that depend on it will be graphed. One can pan around the graph, zoom its size in and out by changing font sizes, hard-copy the graph, and save or load a class file from disk. Each node of the graph is mouse-sensitive, and via various button clicks one can describe, create, move, or edit a class. Similar facilities are provided for manipulating instances and flavors.

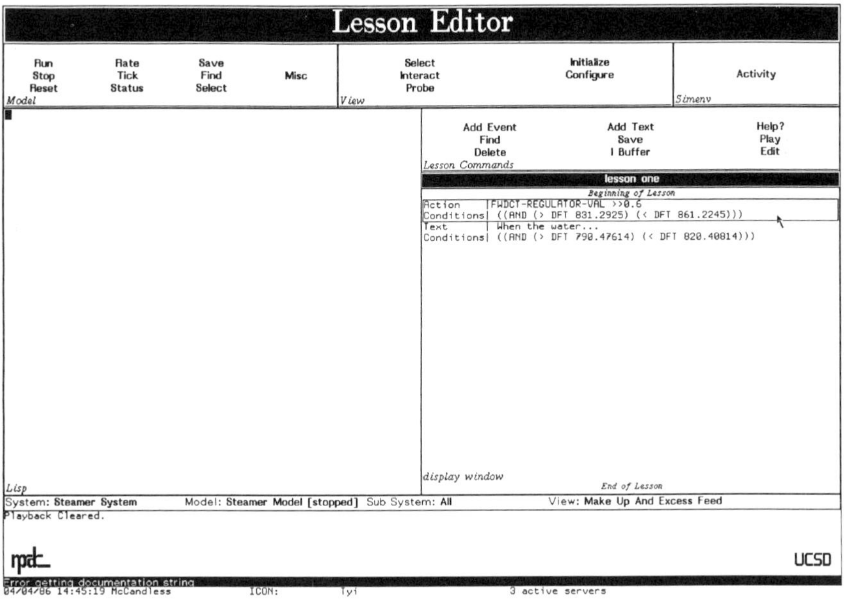

Figure 19. Lesson editor. Screen configuration for control of lesson-editing activity. The result of interacting with a graphic view to specify a portion of a lesson sequence is also shown.

2. Lesson Editor

The *lesson editor* activity provides interface builders the ability to add instructional sequences to a simulation. In the current version of the simulation environment the lesson editor screen appears as in Figure 19. It maintains the full range of model control facilities while supplying additional functions for creating and editing lesson sequences. These sequences are made up of sets of actions each of which are tied to some behavior within the running of a simulation. Each action can either display text or effect the state of the simulation in some way.

The lesson display window shows two actions that have been specified for lesson one. The first indicates that when the model variable DFT (the level of a tank) is in range 831 to 861, the variable FWDCT-REGULATOR-VAL is to be set to 0.6. The second action waits for the DFT to be in range 790 to 820, then presents some text. These actions are specified by first selecting the type of action to add to the lesson (by selecting "Add Event" or "Add Text"). The conditions of activation are then collected by graphically specifying the target states of the desired components within the model. For example, for the actions mentioned above, one would graphically set the desired range for the water level of the DFT by pointing to levels in the iconic depiction (the DFT tank in Figure

4). If the resultant action is to change some component of the simulation, this action is also specified graphically. Text which is to appear as the result of an action is entered in an editor. Once actions are entered into a lesson they appear as condition–action pairs in a menu (in Figure 19 shown under heading "lesson one"). Actions can then be selected, edited, deleted, or moved within the lesson. The lesson can then be saved to a file and placed in a set of lessons for student use.

We currently are in the process of expanding the lesson editor so that a user can employ knowledge which has been provided about simulation objects and views to construct instructional descriptions of components and behaviors within the simulation. We are also designing instructional systems that will gather information about a student's knowledge of a domain by monitoring interactions with the simulation. This system will pose questions to students which are to be answered by interacting with graphic views.

3. *Behavior Editor*

One obstacle to a domain expert being able to use the graphics editor to construct an interface for a domain is the requirement of an existing mathematical model of the domain. Even if a model is available, it is unlikely that a domain expert, without significant additional effort, will have sufficient understanding of the simulation model to be able to tap icons into it. The behavior editor is an initial exploration of a tool which would eliminate the need for an underlying simulation model. The icons in the graphics editor know how to "appear" in order to represent the status of variables to which they are tapped, but the behavior of the system is defined by a simulation model. The behavior editor is composed of icons that know aspects of both the behavior and the appearance of the objects they represent. The behavior of the system emerges from the interactions of the icons with each other. The ability to incorporate the behaviors required in a simulation is facilitated by the object-oriented implementation of icons.

To make intelligent icons capable of generating a simulation, it is necessary to add components with domain and simulation knowledge. We have built a number of icons that know the rudiments of fluid dynamics and understand about connections to other icons. These include, for example, tanks and pipes that know about pressure and flow, so they can be used in fluid systems. Figure 20 depicts a very simple system constructed in exactly the same manner as with the graphics editor, but involving behaviorally "smart" icons that can be connected together. These icons have mechanisms for recognizing when connections should be made, and thus the topology of the view is automatically generated.

An excellent example of the flexibility engendered by icons with connectivity knowledge is the super sensor. This icon is a dial that asks icons it connects to what variables can be monitored and which variable to measure by default. In Figure 20 notice that a super sensor is connected to the left tank to monitor its

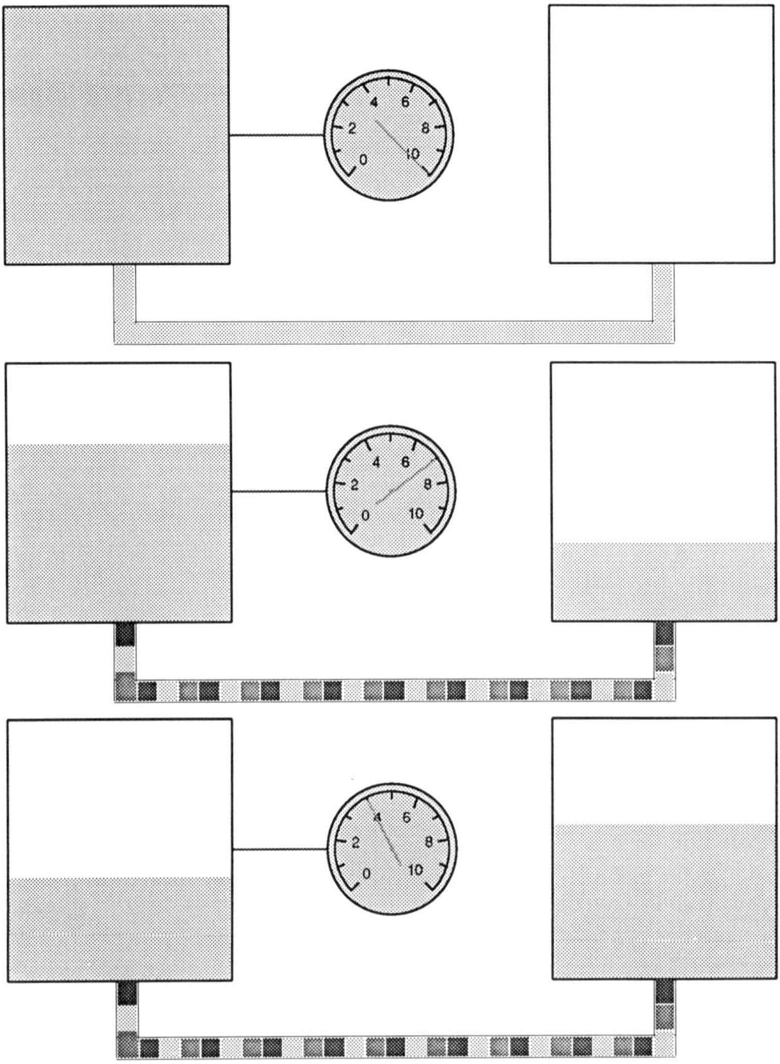

Figure 20. Behavior editor view. Three points in time in the operation of a simple tank and pipe view constructed with the behavior editor. A *super sensor* dial has been connected to one tank and automatically reads the level of the tank.

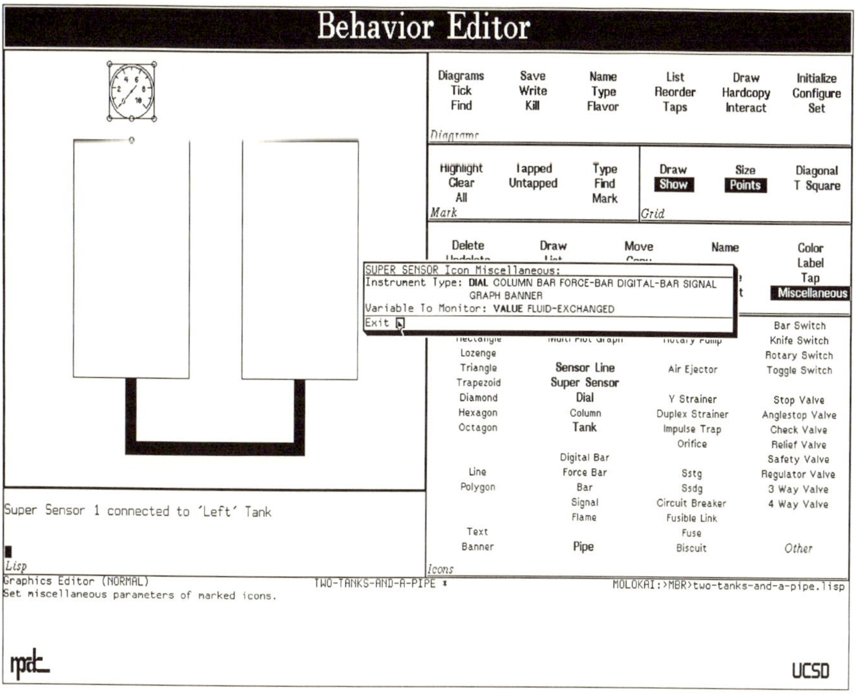

Figure 21. Behavior editor. Screen configuration of the behavior editor during the process of connecting a super sensor to a tank.

level. Sensors know how to monitor appropriate aspects of the objects to which they are connected. Users can manipulate these aspects by clicking on the "*miscellaneous*" menu item on the behavior editor screen (Figure 21). The sensor has defaulted to measure *value,* which is highlighted. The sensor can be changed to measure fluid exchanged by clicking on that menu item. When users interact with the tanks, setting their levels, we are actually setting the initial conditions of the simulation. On the behavior editor screen (Figure 21), they can then click on run, and the simulation will run until equilibrium. It is important to notice a fundamental difference between the behavior editor and the graphics editor. In the graphics editor, the designer would need a mathematical model of the tanks and pipes, and then it would be necessary to find the appropriate variables to tap the icons into. Here the icons themselves perform the required computations for the simulation.

4. *Designer*

Designer evolved out of our development experiences with STEAMER. We found that instructors using the graphics editor sometimes created views that violated simple graphic design rules. They also had difficulty maintaining stylistic conventions across sets of views. Designer is a general tool for assisting with the design process. It functions as an interactive visual design consultant for users of the graphics editor and is intended to aid developers by providing graphic expertise during the construction and critiquing of graphic views. This expertise includes graphic design principles as well as standards of presentation. The underlying motivation is to improve the quality of the views by making them more consistent and visually more effective. In addition to merely describing design alternatives, the system allows exploration of the *design space* by explaining the advantages and disadvantages of alternative design solutions. Through interactive dialogue and constructive examples, the system tutors users of the graphics editor in principles of graphic design.

Designer consists of three interrelated processes, an *analyzer*, a *critiquer*, and a *synthesizer*, coupled to a domain-dependent knowledge base. This knowledge base consists of design elements and relationships, techniques for their identification, and sets of constraints used in distinguishing good design from bad. The analyzer first parses the design based on the elements and relationships of the given domain. The critiquer uses this information to indicate where the current design fails to conform with the principles of good design or predetermined guidelines. Finally, based on searches of the design space, the synthesizer suggests alternative modifications to the current state of the design. The separation of these three processes from the knowledge base provides independence and modularity to the system.

Domain-based design constraints are the basis of the critiquing process. Constraints within Designer consist of both basic graphic design principles important in the construction of two-dimensional views and sets of view standards that are adopted for particular domains. The combination of principles and standards create a context or *style* in which the design critique and subsequent modifications take place. By modifying the style within which a critique is made, one can ultimately affect the form of the final design. It thus becomes possible to request multiple critiques, each based on a different style. This is an especially powerful paradigm for designs that may need to be presented in different media, each with different constraints that need to be considered. For example, a style appropriate for a high-resolution color display may be inappropriate for a black-and-white hard-copy presentation.

An initial implementation of Designer has been completed. A functioning analyzer and critiquer used on existing STEAMER views have provided useful feedback. It is very encouraging that even in views which we had thought were carefully crafted, the system has been able to note inconsistencies and suggest improvements. Progress has been made in identifying the basic elements, relationships, and principles of two-dimensional graphic design and incorporating them into Designer's processes. The analyzer first evaluates design elements in terms of their size, shape, color, and location and then identifies relationships between them from information provided in the knowledge base. These relationships include similarity, proximity, grouping, and repetition. As new relationships are identified, they can easily augment the analysis process. Various techniques exist to interactively inspect the elements and relationships identified within the design.

The critiquer locates examples and violations of the design constraints provided in the current style and creates a critique. These comments include descriptions and justifications based on the graphic constraints from which they were derived. Since the critiquer works within the context of a current style, there are facilities to help define graphic constraints and maintain styles. A preliminary graphic constraint language allows the creation of new constraints, while a style editor has been developed to create, maintain, and switch between styles.

Figure 22 shows Designer's top-level user interface. On the right-hand side are four windows with commands to access views; describe the current state of a view critique; analyze, critique, and synthesize a design; and maintain and edit styles. Status windows attached to these four panes keep track of the current view, the number of elements and relationships analyzed, the state of the critique, and the current style. The middle pane lists examples (+) and violations

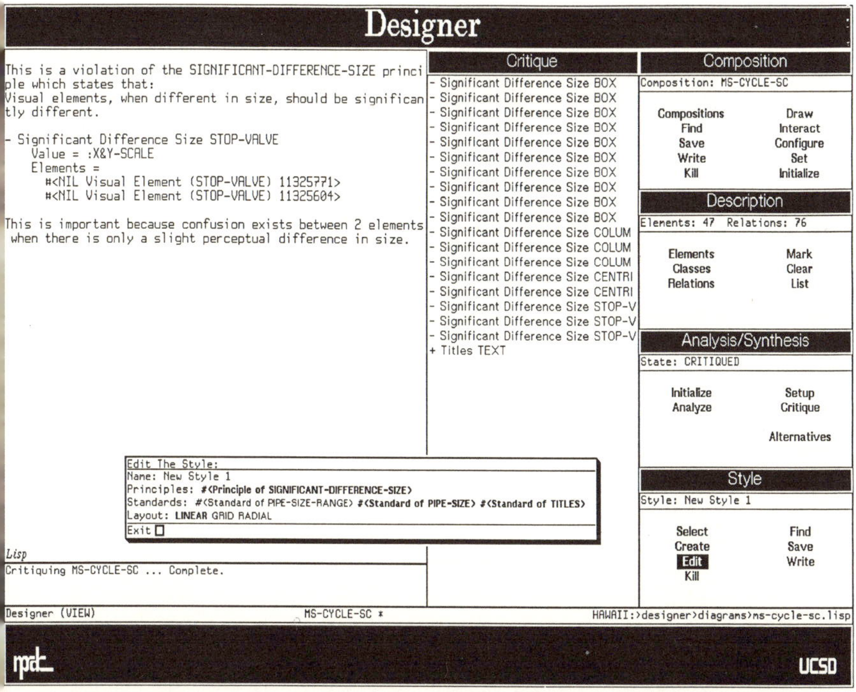

Figure 22. Designer. Screen configuration of Designer.

(−) of the current style's constraints with a more detailed description of one of the violations listed in the larger window pane on the left. In this figure, a pop-up menu is also available to edit the current style, in this case New Style 1.

IV. CONCLUSIONS

Interface design is currently very much more of an art than a science. There is a tremendous need for better theories of interface design and for more powerful tools to assist in their design and implementation. Currently there is virtually no theory of interface design. We do not even understand what contributes to the effectiveness of the most successful interfaces. We are in a state similar to when bridges were built by copying existing bridges without knowing in advance what would result from even the most minor variation. We need a more principled base for the design of interfaces, one that characterizes the dimensions of the space of interfaces. Such a theoretical characterization is the only way to be able to understand and intelligently make the myriad of trade-off decisions inherent in interface design. Hopefully it is clear from this chapter that we think a theory needs to be erected from an understanding of interfaces as representational languages and based on an appreciation of the cognitive tasks that people are employing such representational systems to solve.

One of the factors that influences the development of a theory of interface design is that computer-based interfaces enable new forms of representational languages that are different in fundamental ways from more traditional representational systems. While most representational languages are *static,* these interfaces make possible *dynamic* languages. For example, when we use natural languages or mathematical notation to represent our knowledge about the world, the representational system itself is fundamentally static. The "action" comes from our interpretation of it. Compare this with an interactive graphical interface to a simulation. The representational system *itself* now can "behave," both in terms of reflecting state and in allowing us to directly manipulate it. We still are involved in a process of interpretation but it now concerns behaving entities. The interface becomes a kind of dynamic world in which we can think of and interact with objects as if they were the things themselves. Elsewhere we discuss this very different metaphor for interface design (Hutchins, Hollan, and Norman, 1986). The point here is that this is a novel form of representational system and one which we know very little about.

One of the appeals of these new forms of representational systems are the parallels that exist between them and the forms of representation we employ in perception. When one interacts with the world, the world changes and those differences are reflected in our perception. Interactive interfaces provide a similar form of behavior: we pick up, move, or otherwise modify some object and the associated object in the simulation world changes. As we discussed earlier, this form of representation allows us to employ a number of very effective strategies and to do what we are particularly good at doing: detecting patterns, constructing mental models or simulations of the world that support causal reasoning, and manipulating the world by actions on it or on representations of it. The ability to bring an increasing portion of a dynamic world into the realm of the perceptually knowable is surely one of the major appeals of these new forms of interfaces. It is clear from our experiences developing STEAMER that there is much power from interfaces that provide a form of *conceptual fidelity*. By this we mean interfaces that have characteristics similar to those normally attributed to people's mental models. These include the depiction of state, topology, hierarchical embedding, and the ability to run the models to make predictions about the consequences of changes.

Another conclusion we have reached about graphic interfaces derives from their ability to serve as filters of information. In most of the applications we have built there is an underlying simulation or real-time system. The collection of graphic views that comprise an interface to an underlying system can be conceived of as being a set of filters of information. The designer of a view filters information by selecting what and how to display it. Of particular interest to us has been the ways filtering can be employed to support the development of particular mental models. For example, a message-passing abstraction of a system can be given a graphic instantiation and thus serve to highlight characteristics not normally available and provide an effective way of thinking about the system. Often in STEAMER we have found it advantageous to filter quantitative information and present it qualitatively. One of the very real problems in understanding dynamic systems like propulsion plants is their complexity. A major step in the understanding of process is the isolation of meaningful units to think about and attend to. Much of instruction and the development of expertise is dependent on isolating such units and developing a language that permits talking and reasoning about them. If one looks at the language an expert uses in explaining or predicting a system's behavior, one often sees a restatement of quantitative events in qualitative terms. The filtering of quantitative information into qualitative terms may be an extremely effective means of supporting these types of qualitative explanations and of providing information in forms that encourage development of the mental models needed in reasoning about physical systems.

To build effective interfaces, better tools are required. A major portion of this chapter has been devoted to descriptions of a set of tools we have built to aid in the implementation of interactive graphic interfaces. The goals of our efforts have been to simplify the implementation of interfaces and to make it possible for interface designers to operate at a higher level of abstraction than that normally provided. The object-based graphics editor allows a designer to operate at the level of graphic objects which have been specifically designed for particular domains. It also provides the ability to easily associate these iconic depictions with an underlying simulation or real-time system. The model controller complements the graphics editor by providing an integrated set of facilities for controlling the running of simulation models and interacting with views constructed with the editor. The icon editor increases the generality of this simulation environment by allowing users to construct icons with behaviors that are particularly tailored to the demands of new domains. It thus provides a mechanism for extending the vocabulary of a graphic representational language. We have coupled these tools with a series of knowledge-base editors to allow the incorporation of a wide variety of knowledge. The behavior editor permits the specification of simulation knowledge within graphic icons themselves. The lesson editor makes it possible to build instructional interactions and to make views capable of explaining themselves and their constituents. Designer brings graphic design knowledge to users of the graphics editor. To further augment and support the development of this growing set of tools a general frame-based representational language and graphic interface to it are also available to the interface designer.

All of these software tools have been implemented using the object-based programming techniques of Flavors. There are a number of reasons that an object-oriented paradigm is particularly advantageous for supporting the development of graphic interfaces to simulations and real-time systems. One primary reason is the natural mapping possible between the objects that a simulation models and their graphic representation in an interface. Conceptually this makes possible a natural way of dividing up the simulation world as seen via a graphic interface. Also of principle importance are the relationships that can be made between object-based representations and mental models. The facts that objects store state information, are made of simpler parts, and communicate with and share information with other objects are obvious parallels between the two. These relationships facilitate building interfaces that have some of the characteristics normally attributed to peoples' mental models. In addition, they provide a number of programming features which have proven to be of considerable value. These include nice modularity, the ability to inherit instance variables and messages and thus to easily share common structure and functionality, and ready extensibility by the addition of new messages.

ACKNOWLEDGMENTS

This work is being conducted as part of the research activities of the NPRDC-UCSD Intelligent Systems Group in the Institute for Cognitive Science at the University of California, San Diego. It has been supported by the Navy Personnel Research and Development Center as part of an Office of Naval Technology's Program on Future Technologies for Training. Support has also been provided by contract N00014-85-C-0133, NR 667-541 with the Personnel and Training Research Programs of the Office of Naval Research.

We would like to express our appreciation to David Owen for commenting on a draft of this chapter and to Barbara Morris and Kathy Farrelly for editorial assistance. The opinions expressed in this paper are those of the authors and do not necessarily reflect the views of any government agency.

REFERENCES

Hollan, J. D., Hutchins, E. L., and Weitzman, L. 1984. STEAMER: an interactive inspectable simulation-based training system. *AI Magazine 5* 2:15–27.

Hutchins, E. L., Hollan, J. D., and Norman, D. A. 1986. Direct manipulation interfaces. In D. A. Norman and S. Draper (Eds.), *User Centered System Design: New Perspectives on Human-Computer Interaction.* Hillsdale, NJ: Erlbaum Associates.

Rumelhart, D. E., Smolensky, P., McClelland, J. L., and Hinton, G. E. 1986. Schemata and sequential thought processes in PDP models. In J. L. McClelland, D. E. Rumelhart, and the PDP Research Group (Eds.), *Parallel Distributed Processing: Explorations in the Microstructure of Cognition,* Vol. 2: *Psychological and Biological Models.* Cambridge, MA: MIT Press/Bradford Books.

Shneiderman, B. 1982. The future of interactive systems and the emergence of direct manipulation. *Behavior and Information Technology* 1:237–256.

KNOWLEDGE CAPTURE FOR FAULT-DIAGNOSIS TRAINING

Keith D. Duncan and Nini Praetorius

ABSTRACT

Methods are described which have been employed to elicit verbally stated diagnostic heuristics from the utterances of process operators. Any attempt to access and represent expertise in verbal form will have to contend with (1) distortion of the representation by words, (2) distortion by serial production, and (3) efforts necessarily involved on the part of the expert. Moreover, theories of the acquisition of skill suggest that such difficulties are inevitable and can be anticipated and that any method or technique which overcomes them must therefore be a major research priority.

The development and subsequent use of diagnostic rules in training schemes together provide tests or criteria of how much knowledge has been effectively captured. The two tests consist essentially of (1) the coherence of operators' enquiry when they are "starved" of plant information, and (2) the intelligibility of strategic verbal statements to novice problem solvers.

Advances in Man-Machine Systems Research, vol. 3, pages 165–178

ISBN: 0-89232-659-X

The tests correspond to two major ways in which the development of an expert system can fail: at the stage of "extracting" expertise from an informant, and at the later stage of conveying expertise, in some form or other, to nonexperts. The second test requires that the effectiveness of the knowledge in the system is quantified in terms of how well it enables nonexperts to produce solutions to novel problems. Application of this rather stringent requirement is illustrated in the empirical studies reviewed.

I. INTRODUCTION

A. Research into Fault-Diagnosis Training

Evidence reported elsewhere by the first author and his colleagues shows that training in diagnostic heuristics enables the acquisition of effective and generalizable diagnostic skill, both in the laboratory (Duncan and Shepherd, 1975; Shepherd et al., 1977; Duncan, 1981; Marshall et al., 1981a) and in field studies of operators of a hazardous process (Marshall et al., 1981b).

Although previous reports have described in some detail the training procedures and the levels of diagnostic performance achieved, little attention was paid to the methods employed when developing verbally stated diagnostic heuristics from the utterances of skilled operators in the petrochemical industries. The heuristics so elicited have themselves only been referred to in barest detail.

The account which follows seems timely in view of the growing sophistication of hardware and software vehicles for expert systems, a sophistication which has not been matched by corresponding advances in extracting knowledge from human experts—"the bottleneck" frequently pointed out, notably, by Lenat (1982).

This chapter will therefore discuss the ways in which heuristic diagnostic strategies were elicited and refined, as well as the sometimes different models of plant and equipment held, on the one hand, by designers and, on the other, by operators and users. This leads to a consideration of the status of verbal report in this exercise in applied psychology, not in any ambitious attempt to resolve the controversial issues, but rather to make clear the authors' premises, in particular, when using words like *knowledge, skill,* and *expertise.*

B. Knowledge

We probably have to distinguish at least two sorts of knowledge or expertise:

- Knowledge which is declarative, explicit, reportable; knowledge in the sense of knowledgeable, well informed; knowing *that* something is the case (savoir dire); e.g., which nations fought which battles in the Hundred Years' War; the blood chemistry changes following secretion of steroid hormones by the adrenal cortex
- Knowledge which is implicit, or tacit, not easily reportable; knowing *how*

to do something, skill (savoir faire); procedural knowledge; e.g., regulating the direction and pressure of a cutting tool; the differential diagnosis of viral pneumonia

The terms *skill* and *procedure* may be a source of misunderstanding. *Skill* can be misleading because of sensorimotor connotations, although the adjective *skilled* has long been used of a wide range of activities to denote excellence of performance. *Skill* has the latter sense throughout this chapter. The distinction between declarative and procedural knowledge is probably attributable to Winograd (1974), and cognitive psychology regularly uses procedure as part of the computational metaphor. Procedure unfortunately has older, well-established connotations of invariant, even inflexible, routines which, suffice it to say, is not intended here.

An extensive and lucid exposition of the ''knowing how'' *vs* ''knowing that'' distinction is provided in the classical monograph by Ryle (1949). ''Knowing how'' sometimes, but not always, implies a more general capability than ''knowing that.'' For example, knowing *that* a previously traveled route from the post office leads to the railway station is to be distinguished from knowing *how* to take the most direct, but not previously followed, route from a new starting point to a new destination (Leplat and Pailhous, 1981).

Distinguishing between two sorts of knowledge or expertise usually leads to two closely related observations, namely, the relative accessibility of ''declarative'' knowledge, on the one hand, and on the other, the relative inaccessibility of ''procedural'' knowledge or the representations supporting cognitive skill. Both sorts of knowledge may in some sense apply to the same task. As Berry and Broadbent (1984) have observed, the distinction between explicit or reportable knowledge about a task and performance of that task is increasingly being made in psychology, in the perceptual, memorial, conceptual, and other domains.

Because knowledge or expertise in the second sense of implicit, cognitive skill is as powerful as it is elusive, it will often be the objective, something somehow to be captured by an ambitious expert system. But just how feasible is this and, from the point of view of psychology, how theoretically plausible? In particular, why should expertise in the sense of cognitive skill be relatively inaccessible?

C. Knowledge Capture

A theory of skill to which many psychologists would subscribe, if only for want of a better one, goes roughly as follows. Whenever we discern order, or regular sequences of events, we learn routines. We then learn series of routines, and so on. We continually subordinate and transform information. As our repertoire of routines and subroutines expands, we have correspondingly limited awareness of the various components of our action sequences and of precisely how these achieve our goals. We become less and less aware of more and more.

Even a minimal, noncontroversial theory of this kind predicts formidable problems in knowledge capture. Moreover, when knowledge or expertise is exercised, it is extended and elaborated in the light of success and failure (Hawkins, 1983; Leplat, 1986). It is far from clear to what extent knowledge or expertise, thus refined and enhanced, continues to exist in forms amenable to externalization, that is, amenable to expression by experts in a form intelligible to nonexperts.

Skilled operators may not only have difficulty in the matter of telling someone else how they do what they do. They may also find the process of telling rather tiresome, one which calls for effort. The idea of effort and its avoidance is at least implicit in most versions of the theory of skill. The relegation of initially attention-demanding activity to less demanding subroutines gets a lot of practice. So we should not be surprised if a serious knowledge-capturing expedition *is* an effort. When we persuade our experts to reflect on their expertise, we may be trying to reverse a psychological sequence which runs counter to everyday experience from the cradle to the grave.

At the outset, the potential distortions inherent in the necessarily serial production of protocols were recognized, (Duncan and Shepherd, 1975). We also suspected that much expertise, and probably the most important diagnostic expertise, does not operate in verbal form. If the processes of subordination and transformation of information in the acquisition of expertise are also recognized, we must take seriously the difficulties and the effort of accessing that information by the time the expert has manifestly become an expert.

When, therefore, attempts are made to access and represent expertise in verbal form, we may expect at least three difficulties:

- Distortion of the representation by words
- Distortion by serial production
- Persuading the expert to make the necessary effort

These predictable difficulties are illustrated and confirmed in the studies of elicitation of diagnostic expertise which are now described.

II. STUDIES OF DIAGNOSTIC SKILL

A. The Diagnostic Problem

Using control panel fault arrays of a fictitious industrial process plant (Figure 1), operators and other people working in the process industries were persuaded to attempt to diagnose faults and to talk about the ways in which they went about it. The arrays consist of some 30 indicators and recorders, some of which are control instruments, and an annunciator alarm block. The plant includes unit operations, control loops, and thermal economy systems which produce diag-

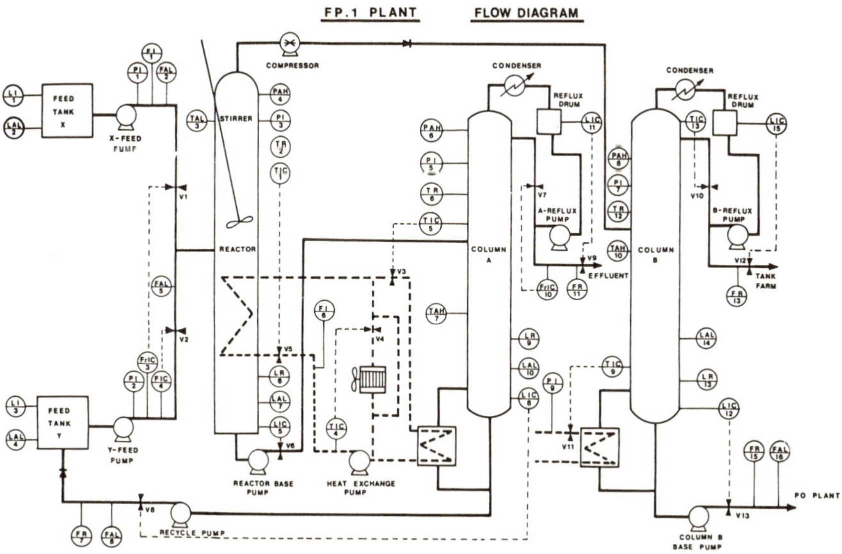

Figure 1. Fictitious chemical process plant (see text).

nostic difficulties typical of chemical process plant, e.g., symptom masking and symptom referral. For further details see Duncan and Shepherd (1975) and Shepherd et al. (1977).

Among those who have attempted the task are "relief operators" from a large petrochemical site comprising several different processes and installations. Relief operators are rather versatile people who are required to go to a variety of plants on the site and take over the control and fault diagnosis problems of an operator who is absent because of sickness or other reasons. If anyone has powerful diagnostic techniques, it seemed reasonable to suppose that these operators should have them.

Over several years, attempts to persuade operators and other people to express how they diagnose faults from control panel arrays show the following:

1. They invariably have difficulty in putting what they are doing into words.
2. They nevertheless can usually comply with what they see as a reasonable request.
3. Such "diagnostic rules" as they seem to be applying are typically short sequences of heuristics.
4. People frequently wish to revise their expressions of these heuristics once they are in verbal form.
5. Further revisions are prompted by further diagnosis and by discussion with other operators in the course of jointly attempting to diagnose a fault.

The information elicited in these verbal reports is subjected to two tests of whether and how verbal expressions may correspond to a diagnostic process.

B. Withholding Plant Information

The first test or check is made possible by a technique of withholding plant information until it is requested. The operator or other subject sees a simulated instrument panel, but without any indications on the instruments. The only information provided is that an audible alarm has occurred, signaling a fault somewhere in the plant. The order in which people request information provides a check of their first attempts at verbalization.

This first test eliminates some tentative diagnostic rules, combines others, and refines still other rules to make them more precise. This is done with individuals and then with groups so that people may compare and debate what it is they are doing, within the acknowledged limits imposed by words, and also with the explicitly acknowledged distortions which this requirement might impose on expressing what they really do when diagnosing a fault in the operating situation.

This procedure helps to turn what could be a rather tedious experience into one which is acceptable and sometimes interesting or amusing. Once the set of heuristics or diagnostic rules seems beyond further improvement—i.e., once further discussion with plant information withheld, in both individual and group sessions, ceases to yield further changes or refinements—the second test is applied.

The first test relies largely on judgments as to whether the sequence of plant information sought is consistent or coherent with verbally reported rules. Sometimes there is little doubt, for instance, that instrument readings are sought in a sequence which is contrary to the rules, however interpreted. Efforts to reduce doubt may be made jointly by the investigator and the informant, or the investigator may resort to judicious choice of faults and therefore of the instrument readings to be sought at some particular juncture. Nevertheless, at this stage there is no attempt to quantify and, to this extent, the first test is fallible. The second test, on the other hand, is essentially quantitative.

C. Enabling Novices

The second test incorporates the verbally formulated diagnostic rules into a training program for novices. This is a test both of how much diagnostic skill has been captured and of how well the verbal expression of this skill in diagnostic rules has been understood and applied by novices.

The population of novices includes both operators and people who would never go into a chemical factory in the ordinary course of events, such as housewives, bus drivers, and students with time on their hands. Figure 2 shows the results of a training experiment in which there were three conditions: no story, where subjects are not told anything about how the plant works; a "theory" condition, where how the plant works is explained, in terms of its

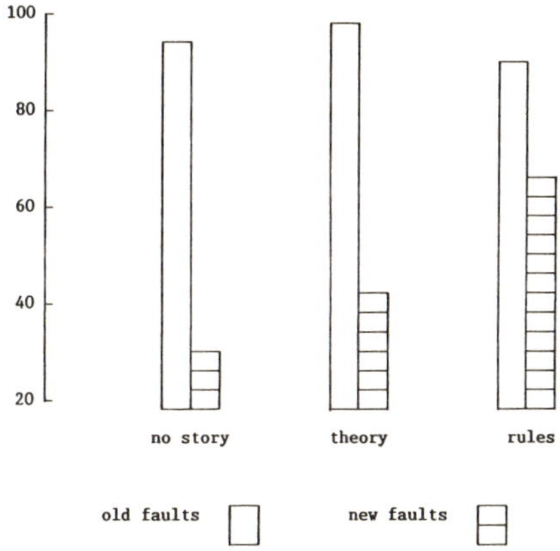

Figure 2. Mean correct diagnoses (percent) of eight old and eight new faults, following three training methods (after Shepherd et al., 1977).

inputs, outputs, the flow between the two, the drives, the control loops, using as simple language as possible; and a third condition in which diagnostic rules are incorporated.

Trainees under each condition were given a post test of 16 failures, 8 of which they had previously encountered during training. A useful but "weak" criterion is the ability to diagnose correctly faults or failures which have been previously encountered. These are referred to as "old" faults in Figure 2, i.e., the panel arrays which had previously been seen during training. A "strong" criterion, which should ideally be applied to any method of diagnosis training, is the extent to which it enables people to cope with faults which they have not previously seen, i.e., with panel arrays, which did *not* occur during training, referred to as "new" faults in Figure 2. It can be seen that the rules condition had a considerable advantage in terms of these novel failures.

The laboratory studies of diagnosis were then extended and validated in collaboration with line managers responsible for the Imperial Chemical Industries (ICI) cyclohexane oxidation process (Marshall et al., 1981b). Cyclohexane is a very volatile hydrocarbon. Its oxidation is part of the route to produce nylon which achieved notoriety, at least in Great Britain, in the Flixborough disaster. In effect, the ICI instructors developed plant-specific training by using as exemplars the techniques and exercises which had earlier been developed in the laboratory.

Figure 3 shows how effective the ICI training program was. The initial low score of experienced operators on novel failures was a surprise. The Test C result is gratifying, but rather what might be expected, since it consists of failures that operators have previously encountered. Test D, on the other hand, shows that ICI operators have achieved rather better results than had been achieved in the

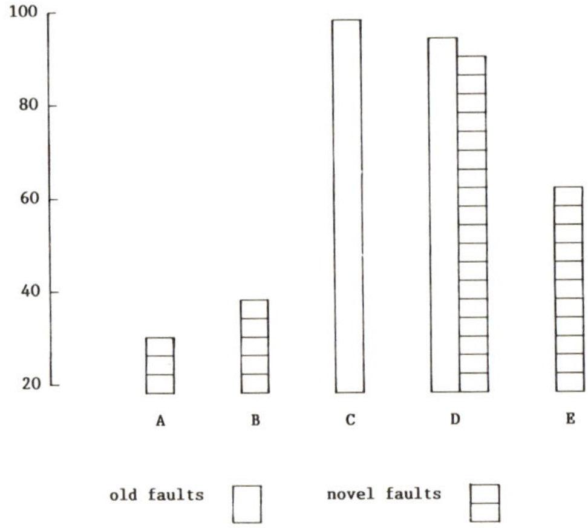

A: initial test, 6 faults
B: test after first account of the process and diagnosis, 6 faults
C: test after practice of diagnostic rules in "withheld mode", 10 faults
D: further post-test, 20 faults
E: final test, 10 "difficult" faults

Figure 3. Mean correct diagnoses (percent) in five successive tests (after Marshall et al., 1981b).

laboratory, i.e., their training has resulted in fault-diagnosis accuracy for novel failures approximating that of failures previously encountered during training.

During the course of this field work, various people, managers, supervisors, and some operators suggested failures, and the indications which would go with them, which they had not considered before the exercise began. They also thought of solutions, or diagnostic rules which might be applied, but which of course had not been included in the training scheme. So these were failures which the rules probably did not cope with at all well, but since, as a spinoff, these failures were now specified, it seemed worthwhile to test trainees on what, by any standards, would be difficult, novel problems (Test F in Figure 3). On these diagnostic tasks, performance was much less accurate, but still substantially better than initial performance, i.e., 66 percent vs. 33 percent before the course was introduced.

III. HEURISTICS, MODELS, AND VERBAL REPORT

A. Heuristics

The laboratory and the field studies together confirm that, while there will be some diagnostic rules which are plant-specific, others may be general to a number of plants. Examples of more general rules would be scanning the panel and considering the possibility of symptoms referred forward or referred back by the process. An effective plant-specific rule applied in diagnosis at the ICI plant was to look for "step changes." A sharp step change on a pen recorder was the target of initial scanning of the panel since, in this pressurized system, it would be very diagnostic of the area in which the fault might be found.

Some of the heuristics which were elicited for fault diagnosis of the fictitious plant (Figure 1) were deceptively simple and straightforward. The first heuristic, "Scan the panel . . .," is one such (see Figure 4). Asked to comment on this rather simple heuristic, operators typically say rather little. One operator said: "You scan the panel for as long as you dare." Another operator who, it seemed, was more able to put into words what "scanning the panel" involved, said, "I know that unless I wait as long as possible before 'chasing that hare' "—referring to one of several diagnoses the operator was considering at the time—"I shall miss the other signs." This seems to be an instance of an operator who has intuitively grasped the well-established human tendency differentially to weight evidence favoring a hypothesis under test and to discount evidence inconsistent with it.

These comments during diagnosis illustrate the shorthand nature of the expressions offered when diagnostic rules are sought and which later exchanges with operators suggests is typical. It underlines how much more a rule may mean than is at first realized. It is also a salutary warning of how much may be lost if one does not persist beyond the first attempts to construct some verbal representation of a diagnostic heuristic.

A related point is that, while it is usually necessary to explain to novices the application of the last two rules which deal with symptom referral, people with experience in the petrochemical industry often spot the fact that built into the

---

1. Scan the panel to locate the general area of failure, i.e., Feed, Reactor/Heat-Exchange complex, Column A or Column B.
2. Check all control loops in the affected area. Are there any anomalous valve positions?
3. High level in a vessel and low flow in associated take-off line indicates either a pump failure or valve failed "closed." If valves OK, then pump failure is probable diagnosis.
4. High temperature and pressure in column head associated with low level in reflux drum indicates overhead condenser failure—provided all pumps and valves are working correctly.
5. If the failure is in the Reactor/Heat-Exchange complex, determine whether it is in the reactor or the heat-exchange system. A failure in the heat-exchange will produce symptoms in Column A but not in B. A failure in the reactor will produce symptoms in both columns.
6. If the failure is in the feed system, check whether it is in stream X or stream Y. Because of the nature of the control system, a failure in the Y stream will produce associated symptoms in both the X and Y streams. A failure in the X stream will show symptoms in the X stream only.

Figure 4. Diagnostic rules (after Shepherd et al., 1977).

fictitious plant is the rather frequent problem of symptoms referred forward or of symptoms referred back by major loops in the overall process.

Heuristics may be evolved because they enable operators to avoid the more laborious procedure of applying their knowledge of process theory when diagnosing faults (Shepherd et al. 1977). On the other hand, Reason (1987) argues that one of the well-established findings of modern cognitive psychology is the overuse of heuristics. Clearly the uncritical use of heuristics could be very dangerous during problematic operations in hazardous industries.

Unfortunately many fault diagnostic problems are not amenable to algorithmic solution because exhaustive specification of the set of possible faults is not a practical proposition. In that case, we may have to resort to heuristic strategies because we have no alternative. *The central problem, therefore, is how to refine heuristics to the point where their success probability is acceptable.* The studies reviewed here show that impressive heuristic strategies for fault diagnosis can be discerned in experienced operators.

However, that said, it must be stressed that training in itself is almost certainly not going to be enough and that we are now reaching diminishing returns. However ingenious the training, the possibility seems remote of improving on first line control room diagnosis of novel failures beyond the 90 percent or mid-90 percent level which can now be achieved. On the strict criterion employed, i.e., problems not previously encountered, this still leaves a high probability of "first-shot" failure, which in many industries is quite unacceptable. The way ahead may be a radical change in how diagnostic problems are represented in the first place—in plant design, before training begins.

B. Representations and Models of Process Plant

Arguing that "technology makes relations between actions and outcomes opaque," Brehmer (1987, pp. 111–112) points out that the user has to contend with the "representations, . . . models created by the designers . . . for the explicit or implicit purpose of facilitating a foreseeable range of decisions," decisions, be it noted, which will not be for the designer to make. We must beware of major differences between the user's representation or model and the designer's model of the system.

There are probably many important differences between the designer's model and the operator's model of process plant. One instance may be the model of a loop. The designer's model of a loop probably has to do with taking an output signal to some comparator and maintaining "set point" by feeding signals to an effector device. At a more detailed level, the designer's model may represent the quality of sensor information, the power and other response characteristics of the slave device, and the comparator functions which ensure adequate tracking of the set point, avoid hunting when perturbations occur, and so forth.

The operator's model of a control loop, precisely because, by its very nature, it copes with plant disorder at least for a time, seems to emphasize the *masking* of process problems in the early stages of their development. In this respect novices differ from experienced operators. Novices usually concentrate on how well the set point is matched, on whether, say, a vessel level indication is within limits. The experienced operator is not unconcerned about matching the set point, but is often very concerned about the behavior of the slave device, say, a slave valve on the feed line to the vessel. If the slave valve, in this instance, tended to remain in a wide-open position, or alternatively to remain nearly closed, an experienced operator would probably infer feed line problems, e.g., changing feed stock characteristics or ''line chokes.''

Developing process problems such as these may be masked, although not indefinitely, by an acceptable match of the parameter sensed (temperature, pressure, etc.) with the set point—and the novice would be no wiser. However, the experienced operator's model keeps him on the lookout for symptom masking and he will anticipate impending problems, provided, that is, that adequate indications of the behavior of slave devices is available. And anticipation is a hallmark of skill.

That control room instrumentation of slave device behavior is often nonexistent or primitive may be taken as further evidence of rather different models held by designers and experienced operators. More prosaically, the quality of control room indications of slave device behavior may merely reflect the financial pressures on designers.

For instance, there are substantial cost differences between displaying the signal sent to a slave valve, displaying ''open'' and ''shut'' signals from limit switches, and displaying the track of the valve position as sensed by an induction device. The instrumentation of Three Mile Island Unit Two may have been the result of an inadequate model of how operators detect developing plant failures or simply the result of pennypinching, or both. In any case, the 1979 accident dramatically illustrates what may be at stake when a design takes insufficient account of the operator's model for anticipating and controlling process malfunctions.

The same point is illustrated by a feature of plant design in the petrochemical industry. It is a consequence of the design engineer's increasing and, it should be said, praiseworthy concern with thermal economy. For example, if a reactor generates heat and requires substantial cooling systems, rather than simply controlling the temperature by cooling towers, the plant designer will often employ heat exchangers and transfer the recovered heat to another plant unit which is driven by heat, e.g., a distillation column.

This makes good sense, but it introduces the diagnostic difficulty that doctors designate as ''referred symptoms'' in medical diagnosis. Thus if a reactor developed a fault, the symptoms might not first appear, and would not only appear, in the reactor instrumentation. The first panel symptoms might be loss of ''boil-up'' in the distillation column.

Do operators always develop a process model that incorporates the ''referred symptoms'' complication? If they do, as some operators known to the first author have, then their model is again different from the designer's, which is primarily a heat-extraction and transport model. In this way another diagnostic problem arises for the operator which, as it follows from other sensible plant design considerations, was probably not envisaged by anyone: the problem of symptoms referred forward or referred back by major loops in the overall process.

Crucially differing models or representations may seem problems enough, but there are probably still other problems rooted in the design engineer's world. Why should a designer consider the multifarious ways in which his ingeniously contrived artefacts will go wrong? The *amour propre* of *some* designers seems to be sophisticated and robust enough for that extended exercise. So what sort of discipline, or training, or experience produces the design engineer who troubles to build in clear indications for every foreseeable predicament of troubleshooting maintenance personnel? These are crucial questions that seem to lead to relatively uncharted territory. Certainly they go well beyond the preserves of applied psychology. Finding the answers may be the bottom line in adapting to the fault diagnosis problems generated by new technologies.

C. Verbal Reports

An important departure from classical introspection, where people attempted to analyze their experiences, sensations, and imagery is the tradition of encouraging individuals to talk, or "think aloud," *while performing a task*. To obtain useful information, eliciting verbal reports in the context of problem solution is of the essence (Miller, et al., 1960, pp. 192–194), a view amply borne out in the empirical studies described in this chapter. Moreover, many of the problems of interpreting subjective reports in the early days of psychology illustrate all too clearly "what happens when 'thought' is analytically divorced from its context of action" (Humphrey, 1963).

In the case of process control tasks, it has been argued, notably by Bainbridge, that protocol obtained from operators provides insights into how they make decisions. She has used protocol material to model the cognitive processes underlying these decisions, acknowledging that it is "necessary to assume that there is some nondistorting mapping from the underlying thought processes to the verbal protocols" (Bainbridge, 1974, p. 148). In this chapter it has been argued that attempts to make such "private" models public may be assessed by the elicitation, development, and subsequent use of diagnostic rules for training novices.

Two tests are proposed as criteria of how much knowledge has been effectively captured. The argument is thus a two-pronged one: for coherence between operators' enquiry and their strategic verbal statements when they are "starved" of plant information, and for the intelligibility of these strategic verbal statements to novice problem solvers.

The two prongs of this argument correspond to two major ways in which the development of an expert system can fail. It can fail at the stage of "extracting" expertise from an informant. Some would argue that failure at this stage is highly likely because humans are unable to "tell" about what they "know" (Nisbett and Wilson, 1977). Other investigators are more optimistic, emphasizing methodological or pragmatic considerations, such as due allowance for the effects of abstraction, or inference, or other distortion from verbal coding, and for the effects of instructions to the subject (Ericsson and Simon, 1980; Leplat and Hoc, 1981).

Development of an expert system can also fail at the later stage of conveying expertise, in some form or other, to nonexperts. In the event, the empirical studies of diagnosis reviewed here show that the much revised diagnostic rules substantially improve the performance of novices on the stringent criterion of novel problems correctly solved.

There were good reasons for expecting that verbal expressions, which in principle might improve performance, might fail to do so in practice, especially when the task involves process control (Crossman and Cooke, 1962; Kragt and Landeweerd, 1975; Broadbent, Fitzgerald, and Broadbent, 1982). However, more recently, in the case of the tasks studied by Broadbent and his associates, it seems that verbal instructions that are carefully managed so as to direct attention and to render critical task features salient may yet prove to be effective (Berry and Broadbent, 1984).

IV. CONCLUSION

The expert must have a nontrivial task in mind. To facilitate the external representation of expertise, the investigator must anticipate difficulties stemming from distortion of the representation by words, distortion by serial production, and efforts necessarily involved on the part of the expert.

Claims of knowledge capture or claims about the "expertise" of an expert system should not be taken very seriously unless the system's representation of expert knowledge is demonstrably communicable. To the extent the knowledge in the system is communicable, its adequacy can be quantified in terms of how effectively it enables people to produce solutions of which they were formerly incapable, and especially to do so in tasks demanding solutions to novel problems. That is the acid test. It is neither an unreasonable test—since expertise is versatile—nor an unrealistic one, as the studies described in this chapter show.

ACKNOWLEDGMENTS

Work by the first author was supported by the Economic and Social Research Council (United Kingdom). The collaboration of the second author was supported by a travel grant from the University of Copenhagen.

REFERENCES

Bainbridge, L. 1974. Analysis of verbal protocols from a process control task. In E. Edwards and F. P. Lees (Eds.), *The Human Operator in Process Control.* London: Taylor & Francis.

Berry, D. C. and Broadbent, D. E. 1984. On the relationship between task performance and the associated verbalizable knowledge. *Quarterly Journal of Experimental Psychology 36:* 209–231.

Brehmer, B. 1987. Development of mental models for decisions in technological systems. In J. Rasmussen, K. D. Duncan, and J. Leplat (Eds.), *New Technology and Human Error.* New York: Wiley.

Broadbent, D. E., Fitzgerald, P., and Broadbent, M. H. P. 1982. Conscious and unconscious judgement in the control of complex systems. British Library experimental electronic journal: *Computer Human Factors.*

Crossman, E. R. F. W. and Cooke, J. E. 1962. Manual control of slow response systems. In E. Edwards and F. P. Lees (Eds.), *The Human Operator in Process Control.* London: Taylor & Francis.

Duncan, K. D. 1981. Training for fault diagnosis in industrial process plant. In J. Rasmussen and W. B. Rouse (Eds.), *Human Detection and Diagnosis of System Failure.* New York: Plenum.

Duncan, K. D. and Shepherd, A., 1975. A simulator and training technique for diagnosing plant failures from control panels. *Ergonomics 18:* 627–641.

Ericsson, K. A. and Simon, H. A. 1980. Verbal reports as data. *Psychological Review 87:* 215–251.

Hawkins, D. 1983. An analysis of expert thinking. *International Journal of Man Machine Studies 18*(1): 48

Humphrey, G. 1963. *Thinking.* New York: Wiley.

Kragt, H. and Landeweerd, J. A. 1975. Mental skills in process control. In E. Edwards and F. P. Lees (Eds.), *The Human Operator in Process Control.* London: Taylor & Francis.

Lenat, D. B. 1982. The nature of heuristics. *Artificial Intelligence 19:* 189–249.

Leplat, J. 1986. The elicitation of expert knowledge. In E. Hollnagel, G. Mancini, and D. Woods (Eds.), *Intelligent Decision Support Systems.* New York: Springer.

Leplat, J. and Hoc, J.-M. 1981. Subsequent verbalization in the study of cognitive processes. *Ergonomics 24:* 743–755.

Leplat, J., and Pailhous, J. 1981. L'acquisition des habilités mentales: la place des techniques. *Le Travail Humain 44:* 275–282.

Marshall, E. C., Duncan, K. D., and Baker, S. M. 1981a. The role of withheld information in the training of process plant fault diagnosis. *Ergonomics 24:* 711–724.

Marshall, E. C., Scanlon, K. E., Shepherd, A., and Duncan, K. D. 1981b. Panel diagnosis training for major-hazard continuous process installations. *The Chemical Engineer* February.

Miller, G. A., Galanter, E., and Pribram, K. H. 1960. *Plans and the Structure of Behavior.* New York: Holt.

Nisbett, R. E. and Wilson, T. D. 1977. Telling more than we know: verbal reports on mental processes. *Psychological Review 84:* 231–259.

Reason, J. 1987. A framework for classifying errors. In J. Rasmussen, K. D. Duncan, and J. Leplat (Eds.), *New Technology and Human Error.* New York: Wiley.

Shepherd, A., Marshall, E. C., Turner, A., and Duncan, K. D. 1977. Diagnosis of plant failures from a control panel: a comparison of three training methods. *Ergonomics 20:* 347–361.

Ryle, G. 1949. *The Concept of Mind.* London: Hutchinson.

Winograd, T. 1974. *Natural Language Understanding.* New York: Academic Press.

APPLICATION OF THE OPTIMAL CONTROL MODEL TO ASSESSMENT OF SIMULATOR EFFECTIVENESS

Sheldon Baron

ABSTRACT

In this chapter, the application of the Optimal Control Model (OCM) to problems related to simulator effectiveness is considered. The discussion is aimed at providing an indication of what has been and can be done to address simulator effectiveness issues with this state-of-the-art human performance model and to illuminate requirements for further research and development. The chapter concentrates largely on simulator fidelity issues because they are of prime concern in device specification, they have a very significant impact on simulator cost, and they are more readily addressed than other effectiveness questions by the available human performance models. In this context, the chapter begins with a discussion of various concepts for simulator fidelity. Then, after a very brief review of the OCM, there is a presentation and discussion of a series of analytical and empirical studies

Advances in Man-Machine Systems Research, vol. 3, pages 179–235.

ISBN: 0-89232-659-X

involving its application to analysis of simulator cuing requirements, analysis of simulator modeling requirements, development of model-based fidelity metrics, and design of compensation for simulator deficiencies. The chapter closes with a brief discussion of some research needs and further potential uses for the model in simulator-training areas.

I. INTRODUCTION

Both the military and civilian segments of aviation are placing an increasing reliance on flight simulators for aircraft research and development and for pilot training and proficiency maintenance. As simulator users and designers have sought more and more capability in the devices, costs for purchasing and operating the simulators have escalated substantially. This increased reliance on, and cost of, simulators and simulation training has raised a number of issues concerning simulator effectiveness. Much of the questioning has centered on the nature of, and need for, "simulator fidelity," inasmuch as increased fidelity is often a driver of costs. Moreover, fidelity is often interpreted as greater physical realism, and there is a lack of convincing scientific proof that such realism is necessary or effective from a training standpoint.

The most extensive and comprehensive investigation of simulator effectiveness was the Simulator Training Requirements and Effectiveness Study (STRES) (Semple, 1981) which examined effectiveness in training in terms of simulator fidelity features, instructional support features, device utilization, and life cycle costs and worth of ownership. A bibliography of abstracts of related research projects was also compiled as part of this study. Most of the work in this field to date has focused on how to measure effectiveness, on data (or lack thereof) concerning one particular cue or another, or on how to improve effectiveness through better device utilization. Very little effort has been aimed at the principal subject of this chapter, which is developing predictive models of the process, including the human subjects, that could be used in trade-off studies by simulator designers or those charged with specifying and procuring simulators.

The main concern here is with problems related to the effective design of simulators for training of, or research associated with, high-value flight control tasks. Examples of such tasks, identified in STRES, are individual and formation takeoff and landing; close formation flight and trail formation, both close and extended; aerobatics; spin, stall, and unusual attitude recognition, prevention and recovery; low-level terrain following; air refueling; air-to-air combat, both guns and missiles; and air-to-ground weapons delivery. In designing a simulator for such tasks, the principal problems[1] are the following:

- Defining and obtaining the data for describing the vehicle and its subsystems
- Design of the cue (motion and visual) environment so as to meet simulation objectives

- Design of the digital simulation model to fulfill the real-time simulation requirements with adequate accuracy.

Defining the appropriate mathematical model and acquiring the necessary data are substantial tasks, but human performance models are of marginal utility in accomplishing them, so this aspect of the problem will not be discussed further.

In specifying the cue environment, the designer must establish the need for particular cues as well as the requisite fidelity for their presentation. The choices made here are important because the validity and utility of the resulting simulation can be critically dependent upon them and because the decisions involve major costs of the simulation. Unfortunately, these decisions are quite difficult to arrive at rationally, inasmuch as the choices depend on complex psychological as well as engineering factors. The requirements will be governed by the purpose of the simulation: in the broad sense, training simulators have different needs than research simulators; more narrowly, whether for training or research, requirements will be problem- or task-dependent (e.g., the need for motion cues in the analysis of aircraft control in a gusty environment will depend on the gust response of the aircraft). Finally, the capabilities of the adaptive human controller both help and compound the problem. The human pilot may be able to compensate for simulator shortcomings and maintain system performance; however, this could result in negative transfer or reduced acceptability of the device in a training environment or an incorrect evaluation in a research simulation.

The design of the simulation model has become increasingly important and difficult as digital computers play a more central role in the simulations. The need for an adequate discrete simulation is also related closely to the cue generation problem inasmuch as the errors and, in particular, the delays introduced by the simulation will be present in the information cues utilized by the pilot.

In this chapter, the application of a particular model for closed-loop control, the Optimal Control Model (OCM), to the above simulator effectiveness problems, and some related ones, is considered. The OCM is a general model for closed-loop control tasks, but here the major focus is on its specialization to simulator problems. Much of the discussion will be couched in terms of training uses of simulators, but the work is relevant to engineering uses of simulators as well. Furthermore, we will concentrate largely on simulator fidelity issues because of their prime importance in device specification and because they are more amenable to treatment by analytic human performance models. In this context, we begin with a discussion of various concepts for simulator fidelity. Then, after a very brief review of the OCM, we discuss a series of analytical and empirical studies involving its application to the following:

- Analysis of simulator cuing requirements
- Analysis of simulator modeling requirements
- Development of model-based fidelity metrics
- Design of compensation for simulator deficiencies

The chapter closes with a brief discussion of some research needs and further potential uses for the model in the simulator-training area to address other questions of simulator effectiveness.

II. SIMULATOR FIDELITY

The concept of simulator fidelity has sometimes received a "bad name," particularly in the training community, because of an overemphasis on simulator realism. However, there are a number of definitions of fidelity, some of which are undoubtedly closely related to simulator training effectiveness. Furthermore, it is likely that any rational simulator design methodology is going to have to rely on some metric of fidelity, but that metric will have to be related to the objectives to be addressed by simulation.

A suitable fidelity metric can provide an intermediate target objective for the design specification process. By associating a specific fidelity *level* with a given objective, one can segment the simulator design process into two phases: an initial phase in which given requirements are translated into a specific set of fidelity requirements (a metric and its associated level) and a second phase in which these fidelity requirements are translated into specific simulator design requirements. The intermediate target objective provided by the fidelity metric ensures a logical and goal-directed simulator specification effort. Cost/benefit analysis can then be factored into such a process by affecting trade-off decisions.

Thus, the definition of an appropriate fidelity metric could be very important to the development of an effective method for translating research and development or research training objectives into simulator design specifications. Here, we briefly discuss some particular fidelity concepts and their potential utility for use in model evaluation of simulators.

A. Physical Fidelity

The simulation engineer has long been driven to achieve a high degree of correspondence between the physical environment of the aircraft cockpit and the cockpit of the corresponding simulator being designed. This goal of *physical fidelity* (sometimes referred to as objective fidelity) provides the engineer with a clear-cut and objective goal which directs the design process and which, if satisfied, is likely to produce a simulator which is well received by the pilot community. Furthermore, it provides a metric which is readily interpretable in terms of specific characteristics of the aircraft and its simulator, and is entirely independent of the less well-defined issues concerning sensory–perceptual processing, human performance, and skill acquisition. In short, physical fidelity provides an exceptionally convenient and reasonably well-understood metric for simulator assessment. Unfortunately, the assessment may be entirely irrelevant to the question of a simulator's effectiveness, particularly with respect to training. It has yet to be convincingly demonstrated that physical fidelity ensures effective training, and the cost of realism can be very high.

B. Perceptual Fidelity

The notion of *perceptual fidelity* concentrates upon the stimulus environment of the pilot and his sensory–perceptual processing capabilities and limitations. The basic notion here is that a simulator need not replicate the physical environment of the aircraft, but it should provide a stimulus set which ensures the replication of the perceptions maintained by the pilot in the actual aircraft. With perceptual equivalence between aircraft and simulator, one can argue that not only is the simulator pilot encouraged to "believe" he is in the actual aircraft (contributing to motivation), but also that having the same cue set available in the simulator as in the aircraft will lead to the same type of control behavior in both situations. The presumption, of course, is that positive transfer is enhanced when the simulator supports the same perceptions and control behaviors occurring during actual aircraft control. Indeed, Heffley et al. (1981, p. 33) have used this notion to formally define simulator fidelity as "the degree to which characteristics of perceivable states induce correct psychomotor and cognitive control strategy for a given task and environment."

Perceptual fidelity implies a comparison between perceptions associated with two different environmental situations: the actual aircraft and the corresponding simulator. We might envision these two systems operating in parallel and acted upon by identical control inputs and environmental disturbances. If the physical stimuli generated by both the aircraft and the simulator were then sensed by identical pilots, and subsequently processed to provide corresponding cue sets, we would then have two perceptual sets, one associated with the aircraft and one with the simulator. The "difference" between these two sets would provide a measure of perceptual infidelity. A perceptually faithful simulation would then be one in which the two cue sets were identical, or at least where the differences between them were sufficiently small so as not to affect overall pilot–vehicle performance.

To design for perceptual fidelity one would require some metric based on the pilot's perception of both the aircraft and the simulator; unfortunately, no direct objective method exists for obtaining this information. We must rely on either subjective pilot evaluations, inferences based on objective measures of performance and/or control behavior, or predictions based on knowledge of sensory–perceptual processing. As is well known, the first is notoriously variable, and the second, though more useful, provides only an overall perceptual–cognitive–motor measure, from which it is difficult or impossible to uniquely infer perceptual processing characteristics. The use of sensory–perceptual concepts and models would thus appear to be a promising method for defining and predicting perceptual fidelity (see Sections III and IV).

The goal of ensuring perceptual fidelity implies that the pilot perceives the simulator state to be practically indistinguishable from that of the aircraft. Were the pilot unencumbered by any perceptual processing limitations, the simulation engineer would be forced to provide the identical stimuli present in the actual aircraft; fortunately, this is not the case, and the engineer has some latitude in matching the two sets of stimuli, with the allowable degree of mismatch a direct function of the pilot's sensory–perceptual capabilities. For example, if it is desired to simulate an aircraft accelerating at, say, 0.01 *g*, one might refer to the vestibular literature to find that such an acceleration is below threshold. Since the pilot in the aircraft would sense nothing, the simulator designer can ensure perceptual fidelity by not accelerating the simulator, a tactic which results in an unfaithful physical simulation, but which capitalizes on our knowledge of sensory–perceptual processing. Clearly, knowledge of other sensory–perceptual capabilities and limitations could be used to similar advantage in generating a necessary and sufficient simulator stimulus set.

With identical perceptions, and the presumptions of identical task objectives and pilot consistency, we would expect the pilot's control actions to be the same under both actual and simulated conditions. If we had a simulator that was not perceptually faithful, we might expect the pilot's control actions to differ between actual and simulated conditions, because he is basing them on two different perceptual sets (presuming identical task objectives and pilot consistency). With a difference in control actions, one may expect the aircraft and the simulator to behave differently, giving rise to a stimulus discordance, which in turn may enhance the perceptual discordance initially present. Thus, rigorous closed-loop analysis could be needed to uncover some of the subtler aspects of stimulus–cue–control interactions in this situation.

C. Performance Fidelity

A concept which is closely related to that of perceptual fidelity is *performance fidelity,* a metric which involves a comparison of pilot-vehicle performance between aircraft and simulator. The relevance of this metric to simulator training effectiveness is discussed by Matheny (1971, p. 39):

> [The] assumption is that the transfer of training between the simulator and the aircraft will be high and positive if the simulator is equivalent to the aircraft in terms of performance requirements on the operator. This means that the measured man–machine system output and the operator performance or input into the simulator controls are not different from those exhibited in the aircraft during the performance of given maneuvers under given conditions. . . .

There are several ways of measuring performance in a man–machine system. For example, in a simple visual tracking task, one might measure the rms tracking error displayed to the human tracker; a more complete description of the system performance would include the rms value of the tracker's control actions (e.g., control stick deflection). If the experiment were suitably configured, one could also make performance measurements in the frequency domain: the human operator's describing function, which directly relates stimulus inputs to control action outputs, and the human operator's remnant function, which accounts for the random or stochastic portion of his control actions. Such measures have had a long history of usage in the area of manual control research and can be directly applied to measuring pilot-vehicle and pilot-simulator performance.

If one were to make the same set of performance measurements in both the aircraft and simulator, then one would be in a position to identify simulator-aircraft differences along each dimension of measured performance. An absence of differences would indicate that the simulator is equivalent (in a performance sense) to the aircraft and that both the simulator and aircraft impose the same stimulus–response pattern on the pilot. Naturally, one would not expect to ever have perfect equivalence between aircraft and simulator on all dimensions of performance, but it would seem likely that there exists some threshold difference level which defines an "effectively equivalent" simulator. That is, once aircraft–simulator performance differences become small enough, one would not expect to see any improvement in a simulator's effectiveness. Any attempt to reduce these differences by increasing the sophistication of the simulator would not be cost-effective in terms of program objectives. Of course, defining this threshold difference level is nontrivial and a subject for research.

D. Error Fidelity

Heffley et al. (1981) suggest that it is important that a simulator induce the same type of errors on the part of the trainee as might be committed while learning to perform the task in flight. They credit Carter (1976) with suggesting that this idea be formalized by using the term *error fidelity* so that a simulator has high error fidelity for a given task when—

1. Students tend to make the same errors in the simulator that they make when learning the same task in the aircraft.
2. The *relative* frequency of the different errors associated with the task in the simulator is approximately the same as that in the aircraft.
3. The effect of each student error on system performance, flight path, and maneuver outcome in the simulator is the same as that in the aircraft.

They claim that error fidelity can have a very significant impact on transfer of training and therefore high error fidelity is an important design goal. As an example, they point out that an error in maneuvering that could result in departure from controlled flight should have the same effect in the simulator as in the aircraft. This example suggests that error fidelity may be considered a subcategory, albeit an important one, of performance fidelity.

It should be noted that error fidelity could also be very important in simulation for system design and evaluation. Clearly, the design engineer wants to induce the kinds of person–machine errors in the simulator that would occur in actual system operation.

E. Workload Fidelity

In addition to distinguishing the perceptual and performance aspects of fidelity, Baron, Pew, and Zacharias (1980) suggest one further dimension of fidelity that should be considered in relation to pilot–simulator behavior; this dimension is *workload fidelity,* or equivalence. The concept of workload is difficult to operationalize, that is, to define in terms of concrete objective measurements. Nevertheless, it seems likely that a simulator that meets the requirements of perceptual and performance equivalence could still be deficient in the sense that it fails to correspond to actual flight in terms of the workload and/or stress that it engenders. To take a specific measurement example, even though the flight and simulator cases produce the same response strategy and performance, the addition of an extra task to be performed simultaneously in the two cases could still produce differential results.

It is perhaps arbitrary whether workload fidelity is defined to be a separate issue from that of performance equivalence or is included to broaden the scope of the definition of what is meant by performance fidelity. However, it *is* important that workload be considered when evaluating simulation fidelity.

III. THE OPTIMAL CONTROL MODEL FOR PILOT–SIMULATOR SYSTEMS

The modeling of human performance has been an extensive and long-standing endeavor, and there is a very large literature on the subject. In this chapter, however, we are concerned with a much smaller subset of the modeling work, in particular, that segment which is of reasonably direct relevance to the simulation and training of flight control and related, tasks. Broader reviews of the modeling literature are readily available to the interested reader (e.g., Pew et al., 1976; Sheridan and Ferrell, 1974; Rouse, 1980).

The most relevant human performance models for flight control tasks and for use in estimating or evaluating simulator effectiveness, at least in the short term, are those from manual control theory. These models are aimed at capturing the "total response" of the pilot from input (cue) to output (control) as well as the closed-loop pilot–vehicle response. As such, they incorporate a number of processes (such as monitoring and state estimation), though the manner in which this is done differs among models and depends on the particular application being considered.

Manual control modeling is dominated by quasilinear describing function models and state-space optimal control models. Both modeling approaches rely heavily on human performance data measured in a steady-state closed-loop control task, with the measurements chosen to succinctly specify the input–output characteristics of the pilot. Typically, these measurements include the pilot's frequency response (gain and phase) over some specified measurement spectrum, along with some composite measure representing the pilot's random, nonlinear, and time-varying behavior (remnant). The frequency response measurements, taken as a complete set over the measurement frequencies, comprise the pilot's describing function, which represents the best linear relationship between the inputs available to the pilot (i.e., those displayed to him) and the outputs generated by the pilot (i.e., his control actions). That portion of the pilot's response not accounted for by the describing function is modeled by the remnant spectrum, a frequency domain description of that part of the pilot's output that is not linearly correlated with the system input (and may often be attributed to a form of pilot "noisiness").

In this chapter we consider only the application to simulator effectiveness issues of the state variable approach as embodied in the Optimal Control Model (OCM); the reader is referred to Heffley et al. (1981) for a relatively recent examination of the application of the describing function approach to some similar issues. Because of the importance of the perceptual cuing environment to simulator fidelity issues, we also review briefly models of sensation–perception that are relevant to manual control tasks and to the OCM.

A. Optimal Control Model for a Pilot–Simulator System

The OCM is a model developed to analyze person–machine system performance in continuous control tasks. The model is more than a model for human control as it contains submodels for perception, attention sharing, and state estimation and has been extended to allow consideration of detection and decision making. The model was introduced in the late 1960s (Baron and Kleinman 1968) and has been elaborated and applied extensively since then. It has been thoroughly documented; the interested reader can find a detailed description of the model, including the underlying mathematical representations in Kleinman, Baron, and Levison (1970) and a relatively recent discussion of various applications of the model in Baron and Levison (1980). Here, we focus on the specialization of the OCM to modeling pilot–simulator systems.

A detailed block diagram of both a simulator "system" and the OCM pilot model is given in Figure 1. Note that the variables in this diagram are generally vectors so that a multiinput, multioutput situation is being modeled. The system portion (outside the dashed box) provides for representations of control stick dynamics, vehicle dynamics, and the dynamics associated with the simulator drive logic (e.g., a motion base washout filter) and its hardware (e.g., the servo drives). As shown, the two inputs to the system are the set of controls generated by the pilot and the system disturbances which perturb the vehicle states. The set of system outputs is the cue set provided by the simulator to the pilot's various sensory systems. The OCM system modeling approach involves (a) a linearization of the relevant dynamics associated with each of the subsystems; and (b) the construction of a state-variable representation of the combined system dynamics.

The state-variable system model includes dynamics associated with all of the

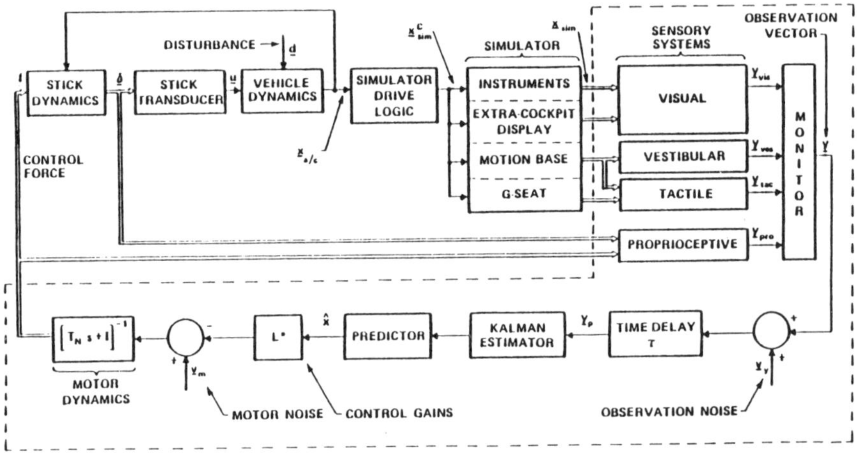

Figure 1. Pilot–simulator model based on Optimal Control Model.

subsystems comprising the simulator in addition to those associated with simulating the aircraft itself and its control and display systems. Furthermore, any dynamics (or state variables) needed to model command or disturbance inputs or to account for pilot sensory capabilities (e.g., vestibular dynamics) are included in the formulation as part of this model.[2] The inclusion of these various elements in the system model is significant because they will then be automatically reflected in the "internal model" of the pilot as represented in the OCM and discussed below. This is tantamount to assuming that the trained pilot will compensate for predictable correlations in the disturbance inputs and for his own sensory limitations, as well as for the particular (simulator) system being controlled.

It is assumed that the flight information available to the pilot is that which he obtains via his visual, vestibular, proprioceptive, and tactile sensory systems. The individual display vectors associated with a particular modality can include information from more than one cuing system. For example, y_{vis} can include both out-the-window visual cues and visual information available from displays within the cockpit.

The pilot cannot attend to all sources of information simultaneously and must therefore selectively allocate his attention. A model for attention sharing has been developed and incorporated in the OCM to provide a basis for predicting the effects of attention sharing on performance (Levison, Elkind, and Ward, 1971). The "monitor" block of Figure 1 represents the attention selection portion of the model, i.e., the mechanism for choosing the fraction of attention associated with the *i*th display variable. (In the steady-state analyses that are conducted most frequently, the overall or average fractions of attention rather than a moment-to-moment allocation of attention is used.) In most studies to date, it has been assumed that the pilot need not share attention between different sensory modalities. This is a plausible assumption but it has not been validated experimentally; to the extent that the assumption is violated, model results would tend to be somewhat optimistic in predicting the benefits of multimodality cues. However, attention may have to be shared among the display variables within a particular sensory modality. This is certainly true for the visual modality and could possibly be the case for other modalities. There are two basic approaches to picking the attention levels, either they are computed to optimize performance or, more simply, they are picked by assumption. When the attentions are picked by assumption, it is most frequently assumed that the pilot devotes equal attention to each display so the attention fraction for each display is the reciprocal of the number of displays.

Additional pilot limitations on the processing of information are accounted for by the time delay and observation noise shown in Figure 1. The time delay τ is a lumped representation of the various internal delays associated with visual, vestibular, central processing, and neuromotor pathways (typically, $\tau = 200$ msec $\pm$ 50 msec). Observation noise v_y is included to account for the pilot's inherent randomness due to random perturbations in human response characteristics and errors in observing displayed variables. A validated model for choosing observation noise covariances exists.[3]

The optimal (Kalman) filter estimator, predictor, and feedback gain matrix represents the set of ''adjustments'' or ''adaptations'' by which the pilot tries to optimize his behavior. The general expressions for these model elements are determined by system dynamics and task objectives according to well-defined mathematical rules (Kleinman, Baron, and Levison, 1970).

The Kalman estimator and predictor are central elements of the model particularly with respect to modeling information processing. Their purpose is to generate the best estimate of the *current* state of the displayed variables, based on the noisy, delayed perceptual information available. These blocks compute the estimate of this state so as to minimize the residual estimation uncertainty. What is being captured is a representation of the subject's ability to construct, from his understanding of the system and his incomplete knowledge of the moment-by-moment state of the system, a set of expectancies concerning the system behavior at the next moment in time. It is these blocks that incorporate explicit ''internal'' models of the dynamics of the system being controlled, the relationships between state and display variables, the statistics of the disturbances driving the system, and the nature of the subject's own limitations.

Given the best estimate of the current system state, the next block (labeled L^*) assigns a set of control gains or weighting factors to the elements of the estimated state, to produce control actions that will minimize the defined performance criterion. As might be expected, the particular choice of the performance criterion determines the weighting factors and thus the effective control law gains.

Just as an observation noise is postulated to account for input processing inadequacies, a motor noise is introduced to account for an inability to generate

Table 1. Features of OCM Relevant to Analysis of Simulator Effectiveness

- CLOSED-LOOP (OPERATOR–MACHINE)
- NORMATIVE/DESCRIPTIVE
- SENSORY/PERCEPTUAL SUBMODELS
- PREDICTION OF OPERATOR ADAPTATION
- MULTIINPUT, MULTIOUTPUT STRUCTURE
- EXPLICIT INTERNAL MODEL
- TIME DOMAIN MODEL
- VARIETY OF PERFORMANCE AND BEHAVIORAL PREDICTIONS

noise-free output control actions. In many applications this noise level is insignificant in comparison to the observation noise, but where very precise control is important to the conditions being analyzed, motor noise can assume greater significance in the model.

Finally, the noisy output is assumed to be filtered or smoothed by the inherent sluggishness of the motor response system. This block, represented in the diagram as $(T_N S + 1)^{-1}$, together with the motor noise, model the limited output capabilities of the human operator.

The most important features of the OCM for application to simulator and/or training analyses are summarized in Table 1. First, we note that the OCM is a closed-loop operator–system model and, as discussed above, will include a representation of the components of the flight simulator, that is, a mathematical description of the simulator design. Thus, as will be seen below, it becomes possible to investigate the impact of various aspects of the simulator design on the combined performance of the human operator and the simulator. For example, various simulator fidelity issues can be explored with the model.

The OCM has been developed for use primarily as a normative model. When standard parameters and assumptions are used in a specific instantiation of the model, it predicts what a highly skilled, highly trained, but inherently limited, operator will do (in terms of performance and control strategy). This has several implications for the analysis of simulator effectiveness. A normative model of this kind provides a prediction of asymptotic training performance as a function of the simulator design. This performance can be compared, for example, with first-flight proficiency requirements to determine whether the simulation fidelity is adequate to meet the training requirement. In addition, normative performance and control strategy provide yardsticks against which one can compare trainee performance during training, i.e., it provides a means for measuring student progress that is related to the ultimate potential of the device. On the other hand, a normative model may not tell us anything about the rate at which skill is acquired as a function of the simulator configuration. Thus, for example, one may be able to use the OCM to predict that motion cues will not be necessary for achievement of criterion performance by a skilled subject while not being able to determine whether or not those cues would be beneficial in the learning of that particular task.

The OCM is not restricted to being used for prediction of performance of highly skilled subjects. Instead, one may "fit" the parameters of the model to existing data. This method of use is essential in the research mode when one is trying to examine, for example, new cuing devices for which human sensory or information-processing parameters have not been determined. It could also prove to be useful for monitoring student progress in training as discussed above. In this case, however, we would describe that progress in terms of information-processing parameters of the model rather than in terms of task performance (or in addition to task performance). This could prove to be more diagnostic of student shortcomings and may provide a better means of adapting the training process than a method based solely on performance decrement.

As can be seen from Figure 1, the pilot–simulator model includes sensory–perceptual sub-models. This feature of the model is important for the examination of the effects of simulator design on perceptual fidelity and performance fidelity. It should be noted that the submodels used in this portion of the OCM are closely related to those proposed by Borah et al. (1978) for use in examining simulator fidelity. However, a key difference is that, here, the sensory–perceptual models are embedded in a closed-loop control model, and this fact increases their utility substantially. Some relevant sensory–perceptual models are discussed in Section III.B.

The parameters of the OCM that relate to sensory processing are observation noise and delay. The observation noise includes a multiplicative component that appears to depend on central processing. This part is characterized by a noise/signal ratio that is a parameter of the model. The observation noise can also incorporate factors to account for display characteristics (such as resolution limits or lack of a reference indication). The delay is a human-related parameter, but is sometimes augmented to account for display delays. There are also parameters in the "output" end of the pilot model, namely a motor noise and a neuromotor time constant associated with each control to be manipulated. The significant point to note is that the pilot-related parameters of baseline observation noise/signal ratio, time delay, motor-noise/signal ratio, and neuromotor time constant are invariant over a substantial range of conditions. This is illustrated in Figure 2 from Levison (1983), where a good match to data for three tasks requiring substantially different control strategies is obtained without any change in model parameters. Note that both the pilot-describing function and remnant spectrum are reproduced. Even better agreement between model predictions and data can be obtained if slight variations in the parameter values with task conditions are made (Kleinman, Baron, and Levison, 1970).

The results in the figure demonstrate another feature of the OCM, namely, that

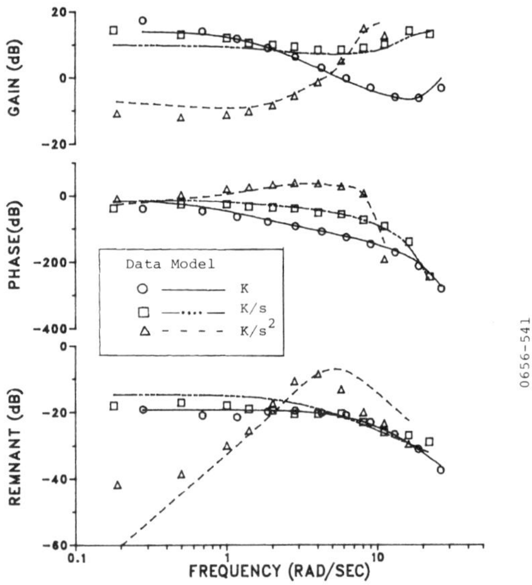

Figure 2. Comparison of fixed-parameter OCM predictions and experimental data for three tracking tasks.

the model predicts the adaptation that the skilled and motivated pilot will exhibit in response to system changes. This adaptation is represented in the model by the monitor, estimator, predictor, and control gain matrix. The ability of the OCM to predict operator adaptation to a variety of task changes has recently been discussed in a paper by Baron (1984).

The OCM provides a predicted multiloop control structure directly once the user specifies an appropriate control criterion (not necessarily an easy task). This is buttressed by the attention allocation submodel, which gives the user a tool for systematically investigating the performance effects of multiple sources of operator information. Finally, the OCM includes operator randomness as an integral part of the multiloop model structure and thus accounts for remnant behavior in an uncontrived manner consistent with the prediction for linear response behavior. These factors make handling of multiinput, multioutput problems with the OCM relatively straightforward.

The OCM assumes the existence of an internal model of the complete task. This model is used implicitly in the computation of the control strategies and explicitly in the Kalman estimator and predictor portions of the model. For predictions of normative behavior, the internal model should be a replica of the actual system configuration and task. However, this is not an absolute constraint on the model, and it is possible to hypothesize alternative internal representations and investigate the resulting behavioral–performance implications (Baron and Berliner, 1977). It is expected that this feature will prove to be important for the study of skill acquisition and, possibly, of individual differences in the learning and performance of flight control tasks.

Finally, we mention two other aspects of the model which are not apparent from the diagram in Figure 1, but are important in terms of the potential for broad application of the model in simulator effectiveness investigations. The first of these is the time domain nature of the OCM, which allows it to be used both for steady-state tasks and for transient or time-varying tasks. This will allow the model to be used to analyze a wide range of specific tasks in flight control. The second feature that derives to some extent from the first and to some extent from the model's inherent structure is the capability of predicting virtually any aspect of performance that can be measured in the simulator as well as some significant "internal" variables that cannot be observed empirically. In the latter category are the state estimates and their error covariances, which may be useful in developing perceptual fidelity metrics using the model (see below). A list of the measures available from the model is given in Table 2.

The optimal control model for the human operator described above is undoubtedly a powerful tool for predicting human performance in continuous control and monitoring tasks. Nevertheless, in its current state of development, there are some limitations and problems associated with its application in the present context that should be discussed.

The problem mentioned most often is the selection of a cost functional. In applying the model to predict performance, it is assumed that the human operator will perform the control task so as to optimize a quadratic functional of the system variables. This functional is defined objectively by specifying relative penalties for various system errors; this specification involves a careful analysis of the problem, say, in terms of task objectives. However, one cannot be certain that human controllers will optimize the objective functional rather than some subjective criterion. Fortunately, experience with the model has revealed few, if any, instances where the choice of performance criteria has actually posed serious difficulties. Nevertheless, the specification of the cost functional remains the primary area involving judgment of the system analyst in applying the OCM.

As we have noted several times, the information processing portion of the OCM includes an estimator and a predictor. These elements contain complete and perfect "internal" models of the system. While we view the explicit internal

Table 2. Predicted Performance–Behavior Measures Available from OCM

- RMS VALUES OF STATE, OUTPUT, CONTROL VARIABLES
- RMS VALUES OF "PILOT" ESTIMATES OF ABOVE VARIABLES
- TIME HISTORIES OF STATE, OUTPUT, CONTROL VARIABLES, AND "PILOT" ESTIMATES OF SAME
- DESCRIBING FUNCTIONS, REMNANT SPECTRA, POWER DENSITIES
- SCANNING STATISTICS
- WORKLOAD

model as a strength or feature of the model, the assumption that the model is perfect appears to present some problems. (Of course, it is only necessary that the operator behave *as though* he has such a model and not that he actually has one). To some extent, model imperfections are accounted for by the observation and (especially) motor noises. Moreover, results to date have indicated that so long as there are random disturbances of any significance, errors introduced by the assumption of a perfect model tend to be negligible. In other circumstances however, e.g., early in training, or for very complex systems, the perfect model assumption may lead to inaccuracies in model predictions.

A related problem concerns the increased order of the pilot model with increases in problem complexity. Modern control theory (and the OCM) treat input shaping and the dynamics of sensors, actuators, or displays by state augmentation. The standard OCM will have a perfect representation for these dynamics and will increase in order accordingly. In addition to raising the question of the operator's internal model, this increases the computational expense for obtaining a solution and suggests the desirability of some reduced-order modeling.

We have seen that a fixed set of pilot-related model parameters can yield a good match to frequency response measures (pilot gain, phase, and remnant) for a significant range of laboratory tracking dynamics, and the match can be improved by "tuning" the parameters slightly for each condition. If we consider a wider range of tasks, however, especially tasks employing dynamics of higher order and/or significant delay, we begin to see that matching the data requires more significant changes in certain "pilot parameters." This could mean that one set of parameters is suitable for wide-band control tasks but that a different set is needed to model slower-responding systems. An alternative explanation would be that these results reflect limitations on the operator's ability to develop or construct an accurate internal model of the task environment in these slowly responding systems. If this is the case, a modification of the adaptive portion of the OCM to include a model of the operator's identification of the system, as suggested by Levison (1983), might account for observed performance without requiring changes in the basic pilot parameters.

Finally, we note a point made by Heffley et al. (1981), namely that the minimization of a performance index does not allow for nonoptimal piloting techniques such as those based on rule-of-thumb procedures. It is true that this area of pilot performance has not been explored to any extent in the OCM. It would appear possible, however, to add such behavior by including appropriate "internally" generated commands that are to be "followed" optimally or by other means, but the specifics of any such approach remain to be defined and validated.

To summarize, the OCM has been used to predict performance for a wide range of conditions. It has been applied to both simple laboratory tasks as well as to more realistic ones (see Section IV). Its formalism allows for extensions to time-varying situations and it can treat multicue environments in a relatively straightforward manner. There are, as with any approach, some shortcomings and some areas needing further research and development. Nonetheless, as will be shown in Section IV, the OCM can be used today to explore several areas of significant concern in analyzing and improving simulator effectiveness.

B. Sensory–Perceptual Models

A critical concern in designing or selecting a simulator is determining what cues are essential for performing and/or learning a given task and how faithfully or accurately those cues must be presented. To address these questions quantitatively requires models for human sensory–perceptual mechanisms. Below we discuss some of the existing models and their relevance to the OCM.

1. Informational Analysis

Before discussing sensory modeling, per se, it should be noted that for analyzing many control tasks it may not be necessary to have detailed models for sensory mechanisms. Instead, one may be able to account for the phenomena of interest via an informational approach in which variables that are needed for control and are presumably available from the cue environment are assumed to be provided as basic "display" information. This information can be less than perfect (being corrupted by perceptual noise, thresholds, etc.), but the mechanisms that transduce stimuli are not explicitly modeled. The informational approach is generally valid when the "sensed" signals are such that human sensory dynamic limitations do not come into play in a significant manner (e.g., when motion stimuli are within the passband of the vestibular system). This is also a particularly useful approach for cases where adequate sensory models do not exist in that it provides a method for estimating a "bound" on the potential improvement to be gained from including certain informational cues.

Informational analysis has been used successfully with both classical and optimal control models to predict and describe performance in a variety of control situations. For example, in modeling the multiloop automobile driver, McRuer et al. (1973) use heading and lateral position as feedbacks for control, though they noted that this in no way implied direct perception of these variables. Levison and Junker (1977) were able to predict the effects on performance of roll-axis motion cues on the basis of such an informational approach while Baron (1981a, b) used it to predict the effects of *g*-seat cues (see Section IV below).

2. Visual Perception

The most important input for flight control tasks is generally the visual input. Major simulation hardware and software are devoted to reproducing cockpit instrumentation and to providing external visual field information where that is appropriate. The modeling of the use of standard cockpit instruments by the pilot is relatively straightforward, with the exception of the modeling of the visual scanning process. The situation for pilot perception of external visual cues is by contrast much more complex to model. It is also significantly more difficult to provide adequate cues for the external visual scene, and this is a subject of much effort in the simulation field.

The literature on human visual perception is extensive, but it is not clear that much of it is of help in providing models for inclusion in pilot–simulator models for closed-loop control. The critical issue for modeling of visual information processing is not so much the characteristics of the visual sensors as transducers as is the case with other aspects of sensation. Many aspects limiting the visual system such as resolution, saturations, and so forth are fairly well understood and have been measured. Instead, the difficult questions revolve around what information is being used by the pilot in performing tasks that rely on external visual field information.

Initial approaches to visual perception modeling in the classic manual control models were primarily informational. In particular, for purposes of analysis, it was generally assumed that the quantity needed for feedback control was available directly. McRuer and Krendel (1974, p. 65) put it thus with respect to multiloop describing function theory:

> The inputs sensed for VFR conditions are currently estimated on the basis of control needs (i.e., what feedback paths are necessary or desirable for the closed-loop system). The actual quantities perceived are likely to be linear combinations of these with the weightings between the inputs fixed by the geometry and perspective rather than being independently adjustable by the pilot. These aspects of perception can have profound effects on the closed-loop analysis of various maneuvers.

Recently, there have been efforts aimed at including some of the geometric and textural visual cues as information sources for pilot describing function modeling. Heffley et al (1981) include functions for perception of height, visual range, lateral deviation, etc.; their report contains a fairly extended discussion of the sources of various visual cues and of their relation to vehicle control techniques. The functional relations are similar to those developed earlier by Wewerinke (1979), Grunwald and Merhav (1974), and Baron, Lancraft, and Zacharias (1980) for use in the OCM.

Borah et al. (1978) used an informational approach for considering peripheral visual motion cues that might be provided by the passage of stars or clouds in a wide-field simulation. Their principal concern was for visually induced effects on orientation and motion perception, generally referred to as "vection." The self-motion perception provided by this cue was limited by visual thresholds and observation noise.

There have been several efforts aimed at including external visual cues in the OCM. These studies have ranged from basic laboratory investigations of the use of such cues to more applied studies in which the impact of changes in computer-generated image system parameters on performance were predicted. All of the work thus far has been concerned with low-altitude flight tasks such as approach-to-landing, hover or following a roadway. In all the studies to date, there has been no attempt to describe and quantify the full stimulus environment provided by the extra-cockpit scene. Instead, the modeling effort has centered on defining those cues that are directly useful for controlling the aircraft. This has led to a concentration on the information provided by geometric visual cues and, secondarily, on that provided by textural cues. Essentially all the studies of the use of external visual cues for control tasks follow the general approach for including visual scene information in the OCM outlined by Baron and Berliner (1977). For some of the later developments in visual scene analysis methodology as it pertains to the OCM, as well as for a thorough discussion of the basis for that methodology, the reader is referred to the report by Zacharias and Levison (1980).

Perhaps the most elaborate visual cuing model developed thus far for the application to simulator design is that of Zacharias, Caglayan, and Sinacori (1983). They have developed a computational model for visual flow field cuing and self-motion estimation based on the notion that the moving observer makes noisy sampled measurements on the spatially distributed flow field surrounding him. On the basis of these measurements the observer is assumed to generate an estimate of his own linear and angular velocity which satisfies optimally, in a least-squared sense, the visual kinematic flow constraints. The underlying cues are provided by changes in lines of sight to viewed objects which result from the dynamically changing geometry between the observer and the object. Zacharias et al. have used the model to simulate visual performance for a pilot flying straight and level over a flat, featureless, textured terrain. They have shown that the model is sensitive to variations in display parameters such as out-the-window field of view or number of points in the viewing region. Experiments are being conducted to validate their results.

The incorporation of this visual cuing model in a closed-loop pilot simulator model is not a trivial matter. One reason is that the visual perception model itself is quite computationally intensive and, at this time, has to be run separately. This is very different from the situation with the other sensory perceptual models considered in this section. Thus, it is necessary to develop an appropriate interface between this visual perception model and the OCM. Levison, Zacharias, and Sinacori (1982) presented some early results using an approach in which the perceptual model was "run" to generate error statistics which were then related to observation noises in the OCM. Experimental programs to validate this approach to interfacing the models are also underway.

3. *Vestibular Sensors*

Motion cues and their relation to piloting and to flight simulation have been the subject of considerable research. Most of the early efforts were aimed at cataloging the effects of whole-body motion on pilot tracking and maneuvering behavior, rather than at the development of models for predicting the effects of

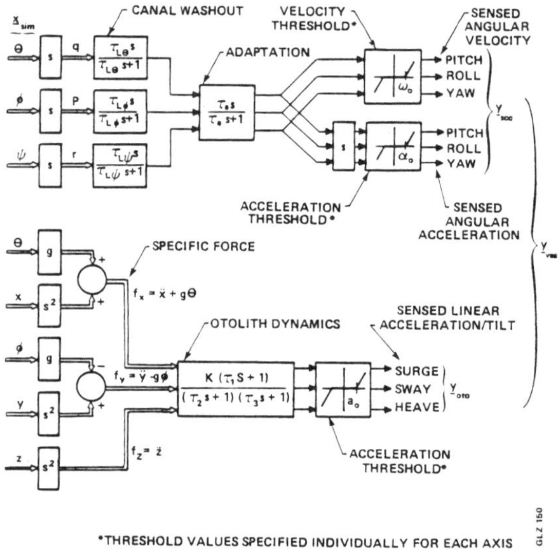

Figure 3. Model for vestibular system dynamics.

motion cues on performance and learning. There have been several reviews of the motion cuing literature (e.g., Borah, Young, and Curry, 1978, and Zacharias, 1978) and, based on them, models have been postulated for the vestibular sensing organs that would be useful for pilot–simulator modeling. These models are very similar, as might be expected. The one proposed by Zacharias has been incorporated in the OCM and used to analyze motion cue effects, so discussion here is limited to that model.

Figure 3 shows Zacharias's vestibular model in block diagram form. (Specific model parameter values, taken from Zacharias's review, are given in Baron, Lancraft, and Zacharias, 1980). The upper portion models the semicircular canals as transducers of angular velocity of the head, while the lower portion models the otoliths as transducers of specific force (linear acceleration minus gravity). These signals, in conjunction with a gravity model and adequate computational capability (assumed to be provided by higher-level processing), can provide a complete specification of body orientation and position in space, along with the angular and linear velocity derivatives. Thus, the vestibular system is often viewed as the primary sensory system for motion sensation. Nonlinear threshold functions in the model provide a means of characterizing the pilot's perceptual limitations in sensing motion.

4. *Tactile Sensors*

Recently, there has been increasing interest in modeling the pilot's use of tactile cues as devices such as *g*-seats, *g*-suits, and helmet loaders are considered for simulators, particularly as substitutes for more expensive motion platforms.

Borah et al. (1978) reviewed the sensory literature and proposed two similar, relatively simple models for tactile sensation that include a model for body seat compression dynamics based on the work of Gum (1973). These tactile sensation models are assumed to be stimulated by a specific force input representative of that in flight. However, in a simulator the signals used to drive a tactile cuing device are at the disposal of the designer and may be chosen to differ from what is expected in flight.

Levison, McMillan, and Martin (1984) also examined the issue of modeling tactile sensation specifically as it relates to the operator's use of *g*-seat cues. They performed closed-loop sensitivity analyses with the OCM and determined that mechanisms were needed to account for both low-quality, zero-derivative information and high-quality, first-derivative information about the *g*-seat drive signal. Based on a review of the relevant physiological and psychological literature, they proposed a simple lead-lag mechanoreceptor transduction model for inclusion in the OCM to account for cues provided via a *g*-seat. The model is embedded in an overall model for the simulator that includes equations for the drive algorithm and for the dynamics of the *g*-seat. The lead-lag receptor model of Levison is of the same form as that used by Borah et al., but Levison uses different parameter values.

5. *Proprioceptive Sensors*

The modeling of proprioceptive sensors is also in a relatively primitive state. Moreover, Borah et al. (1978) noted that it is a difficult sensory mode to approach because it involves many interrelated joints and muscles, and afferent signals available to the central nervous system are not well defined. The combination of all the proprioceptive senses permits subjects to perceive body accelerations.

Borah et al. included one source of proprioceptive sensing in their sensory model. In particular, they considered muscle spindle response to lateral head motion with respect to the body trunk. Their model for proprioception was based on a model for the head–neck control system developed by Gum (1973). Baron (1980) considered the inclusion of proprioceptive and kinesthetic cues available from a control loading system. An informational approach was employed in which the proprioceptive cues were assumed to provide *observations* of stick position and stick rate while the kinesthetic cues provided observations of stick force. This work must be considered to be of a very preliminary nature.

We should also mention the detailed neuromuscular dynamic models developed in support of describing function pilot models and for biodynamic modeling. For the most part, these models have not been used in connection with proprioceptive sensory functions, although such may be possible (e.g., see Heffley et al, 1981).

6. *Unified Multisensory Modeling*

For most aspects of simulator perceptual and/or performance fidelity, one is concerned not with a single sensory mode but rather with the complex issues

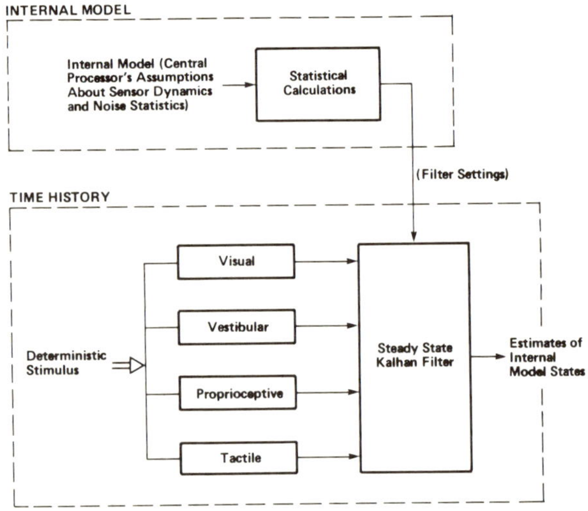

Figure 4. Kalman filter sensory blending model (from Borah et al., 1978).

associated with a multicue environment (e.g., cue substitution and cue conflict). This requires integrating all the submodels into a unified model for sensory processing, which in turn requires modeling higher-level processing as well as sensory transduction. Two closely related approaches aimed at this integration are the unified sensor model of Borah et al. (1978) and the multicue versions of the OCM (Baron, 1980). Both rely heavily on optimal estimation theory, in particular Kalman filter theory, to account for the higher-level processing and, thereby, cue integration.

Figure 4 from Borah et al. shows the basic structure of their motion and orientation perception model, where the estimation process is based on a Kalman filter. Extensive computer simulation was conducted with this basic model to explore its behavior and to "tune" the parameters associated with the estimation process. The model produced results that were in good agreement with a large number of the characteristics of human motion perception (e.g., circularvection and linearvection responses) with only a couple of unreasonable predictions (a forward acceleration step in the "dark" failed to produce an appropriate velocity estimate and tilt motions resulted in a small but puzzling prediction of translatory motion sensation).

Borah et al. also suggested some nonlinear additions to the model. Perhaps the most interesting and significant of these was a cue conflict resolution scheme based on the model of Zacharias and Young (1977). They also suggested means for accounting for the active pilot, noting the importance of mental set for perception. However, they did not consider the active pilot in a closed-loop environment.

The multicue OCM does embed a unified sensory model in a closed-loop control environment (see Figure 1). Thus, the active pilot is accounted for directly with the Kalman filter obtaining knowledge of both control inputs and expected vehicle response. In addition, inclusion in the OCM allows attention-sharing and workload effects to be modeled, and these can have a significant impact on perception. Accounting for all of these factors is essential for performance prediction and, in the author's view, also for evaluating perceptual fidelity as it might be experienced in situ. Moreover, there does not appear to be any real penalty for considering this total closed-loop situation.

IV. APPLICATIONS OF THE OCM TO SIMULATOR FIDELITY ISSUES

In this section, we review applications of the OCM to the following problems:

- Analysis of simulator cuing requirements
- Analysis of simulator modeling requirements
- Development of model-based fidelity metrics
- Design of compensation for deficiencies in simulator fidelity

These applications relate directly to simulator fidelity questions and constitute an important subset of the simulator effectiveness issues relevant to flight control research and to training flight control skills. The discussion contained herein is designed primarily to illustrate how the model may be used to address these issues. As part of the discussion, however, results of some of the applications will also be presented.

A. Analysis of Simulator Cuing Requirements

To apply the OCM to problems of cue requirements and fidelity, one must specify the existing (or potential) cue environment in terms that are relevant to the model. Essentially, this involves determining what cues are (or will be) available to the subject pilot and how those cues are related to the underlying task variables. This process has been described at some length in the references (see, e.g., Baron, Lancraft, and Zacharias, 1980).

Briefly, referring to Figure 1, it is necessary to determine how the simulator and human sensing and information-processing mechanisms mediate the relations between aircraft state variables ($x_{a/c}$) and human "perceived" variables (y_p). This is accomplished in three steps: (1) mathematical specification of the simulator cuing mechanisms, i.e., of the transformations relating x_{sim} to $x_{a/c}$; (2) mathematical modeling of human sensory dynamics, i.e., of the transformations between y_{vis}, y_{ves}, and so on and x_{sim}; and (3) specification of human perceptual limitations in terms relevant to the task description, i.e., of the values of v_y and other related model parameters. The first step is intimately related to the availability of design specifications for the simulator, whereas the second relies on human engineering models for sensory mechanisms. The final part of this task, the specification of human perceptual limitations in terms of model parameters, depends upon the availability of appropriate model-related data for the particular sensory process involved; these data can be, and are being, obtained from properly focused laboratory research. The entire cue analysis and definition task can be very straightforward, as is the case for cockpit instrument cues and, to a somewhat lesser extent, for platform motion cues; or, it can be very difficult and even not completely understood, as is the case for extra–cockpit visual cues and for certain tactile and proprioceptive cues.

In the rest of this section, we discuss several studies in which the model was used for the analysis and/or establishment of cuing requirements.

1. *Motion Cues*

The need for motion cues in flight simulators has been a subject of much debate in recent years. Although pilots invariably prefer the richer, more realistic, cue environment provided by well-designed platform motion systems, the actual improvement of performance or training effectiveness that results from incorporating these expensive devices has been questioned.

There are a number of ways in which pilot performance may be influenced—either beneficially or detrimentally—by motions cues[4] encountered in real flight. Five broad categories of motion effects are mentioned by McRuer and Krendel (1974). The first three categories consider the alerting aspects of motion which tend, in effect, to motivate the pilot to work harder at his control task (or, in the case of imminent danger, initiate a preprogrammed recovery maneuver). The remaining two categories consider the beneficial use that the pilot makes of sensed vehicle motions in performing his control task and the potentially adverse effect of illusions related to misinterpretation of motion cues.

A fundamental problem in determining the requirement for motion cues for a particular simulation is that there is no universal, simple set of rules for predicting the impact of cues, although some general indications of when simulator motion is important have been identified from empirical research. For example, in his review on simulator training effectiveness, Caro (1979) discusses the influence of motion cues on simulator pilot performance. He distinguishes between "maneuver" motion (that generated by the pilot himself) and "disturbance" motion (that generated by external influences such as gusts or engine failures), and notes the following (p. 495):

> platform motion appears to produce quicker, more accurate simulator control because the disturbance component of that motion resulting from simulated turbulence or equipment failure can provide more rapid and relevant alerting cues about forces acting upon the aircraft than could be obtained from other cue sources. Maneuver motion does not fulfill an alerting function, because it results from pilot-initiated control movements. Research involving maneuver motion . . . suggested that this component of platform motion has little effect upon the control of an aircraft whose flight dynamics are stable. For unstable vehicles, however, the presence of maneuver motion will allow the pilot to maintain control even in flight regions where control by visual cues would be impossible. Thus, disturbance motion permits more rapid and accurate aircraft control under all flight conditions in which such motion is appropriate. Maneuver motion is a useful control cue only when the aircraft is unstable.

This is a broad generalization that appears to be largely correct with respect to system performance. However, it has been seen that pilot strategies, as measured by describing functions, are affected by motion cues in both the disturbance and maneuver motion cases (see below). Moreover, the statement does not provide any information concerning the magnitude of the effects of motion in any situation.

In the remainder of this section, we show how the OCM confirms and extends the above insights by examining a few applications of the model that explore the effects of motion cues. The results demonstrate the utility of the pilot–simulator model in establishing when motion cues are important for pilot performance. The first application involves using the OCM to predict the effects of motion cues on performance of a Harrier aircraft in different flight control tasks and with different vehicle and control configurations. The results clearly demonstrate how the importance of motion cues can vary dramatically with the specifics of the task. The second application involves using the OCM to model data obtained in basic research into motion cuing. The results show the model to be both a highly accurate predictor of the effects of motion and a unifying construct for the apparent differences in these effects. The final application discussed in this section compares model predictions with experimental data for a complex simulated task and shows the OCM is capable of handling much more complicated simulator environments.

Baron (1981a) conducted an analytical investigation of pilot control of a "simulated" AV-8B (Harrier) aircraft using the OCM. The Harrier is an advanced aircraft capable of both a high-speed forward flight and vertical takeoff–landing and hover. From the standpoint of this discussion, the study was noteworthy in that widely disparate flight control tasks were analyzed as a function of three different simulator motion cuing conditions. In addition, the interaction between these conditions and the level of pilot attention and/or skill (or training) was investigated by means of a sensitivity analysis in which model parameters of the OCM related to human information processing were systematically varied.

The flight control tasks considered were maintenance of hover position in turbulence and attitude regulation in high-speed cruise. The number of actual conditions investigated was doubled by considering, in each instance, the effect of having the Harrier's stability augmentation system (SAS) operative (SAS-ON) or inoperative (SAS-OFF). For all conditions investigated, a severe disturbance gust was assumed to be acting on the vehicle, so all conditions analyzed corresponded to a disturbance motion situation as described above. The three motion-cuing conditions analyzed corresponded to a fixed-base simulation, a simulation including platform motion, and a simulation in which a *g*-seat was assumed to provide cues for surge motions and for pitch and roll angular velocities and accelerations. A detailed description of all the relevant factors can be found in the cited reference.

Figures 5 and 6 summarize the key results of the analysis. In Figure 5, rms

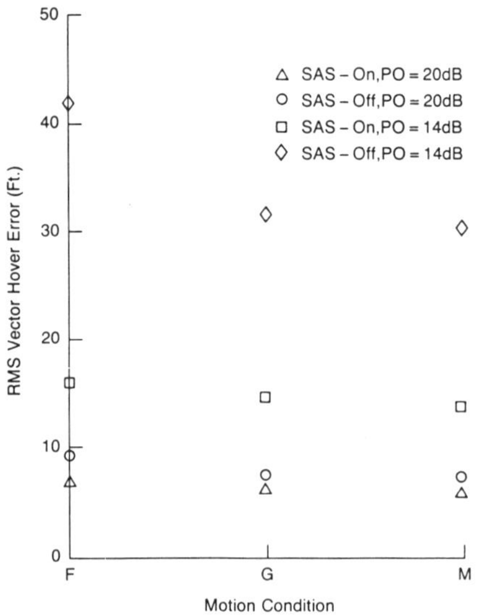

Figure 5. Effect of motion condition on Harrier hover performance.

vector hover error is plotted as a function of the assumed motion condition (F, G, and M refer to fixed-base, g-seat cuing, and platform motion conditions, respectively). The hover error is the square root of the sums of the squares of the three Cartesian components of positional error with respect to a desired hover position located at a fixed altitude over a point on the ground (i.e., error equals the $\sqrt{x^2 + y^2 + z^2}$). For Figure 6, the vector angular error is similarly defined as the square root of the sums of the squares of the individual attitude errors associated with the attitude regulation task for cruise (i.e., angular error equals $\sqrt{\phi^2 + \theta^2 + \psi^2}$). In each figure, results are presented for both the SAS-ON and SAS-OFF conditions. In addition, the base observation noise/signal ratio P_0 of the OCM is used as a parameter. This parameter relates to both level of attention devoted to the task and the level of skill or training of the operator. A value of P_0 equal to -20 dB corresponds to the average observation noise/signal ratio measured in single-axis laboratory tracking tasks in which subjects are highly practiced and well motivated. A value of P_0 equal to -14 dB could correspond to a case in which the pilot had not been fully trained and/or to one in which a highly skilled pilot had other tasks to perform while attempting to control the vehicle in the presence of the disturbances.

The results in Figures 5 and 6 show that motion cues are of significant potential benefit to performance in only one case namely for the hover control task with the unaugmented vehicle (SAS-OFF). In all other cases, motion cues are of minor benefit with respect to performance, with the predicted improvements probably lying in the range of inter- and intrasubject variability. This lack of a significant motion effect obtains, despite the fact that all cases correspond to

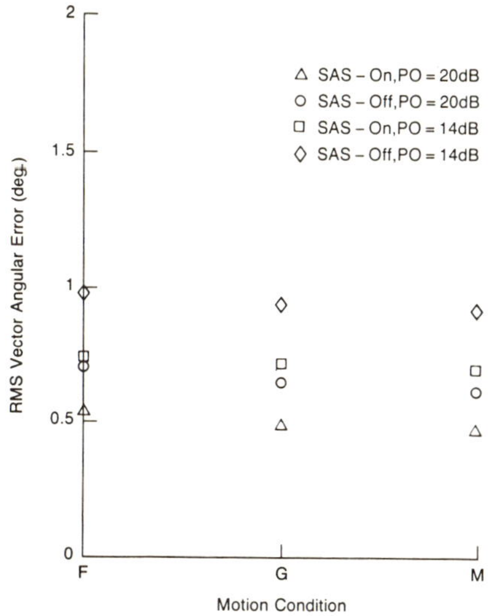

Figure 6. Effect of motion condition on Harrier cruise performance.

a disturbance motion condition (In addition, as will be seen later, motion cues have significant benefit in hover control tasks with different vehicles.)

The model analyses do show that motion cues could be very significant in the Harrier hover control task for the unaugmented vehicle (SAS-OFF) at high observation noise/signal ratios (i.e., $P_0 \leq -14$ dB). This suggests that motion cues would be most important in early training of hover (low skill levels) or when the pilot has other tasks to perform (e.g., weapons control) while attempting to hover, as might be the case, for example, in a full mission simulator. For the SAS-ON configuration, the improvement provided by motion cues is marginal, so the prime determinant of the need for motion cues in this case is the vehicle dynamics (i.e., how easy the airplane is to fly in that condition).

The model also predicted that the addition of motion cues for the task of disturbance regulation in cruise would not result in a significant (i.e., greater than 15 percent) improvement in performance for either SAS-ON or SAS-OFF configurations (see Figure 6). The result held, moreover, for the range of observation noise/signal ratios considered. Indeed, for cruise, the more skilled or the more attentive to the task the pilot (i.e., the smaller the observation noise), the more beneficial the motion cues were for performance. These results, showing the relative unimportance of motion cues in cruise, help to explain why pilots often consider a motion platform to be unnecessary for this flight condition.

An important result of the analysis was that provision of motion cues via a *g*-seat yielded performance approaching that obtained with platform motion. Actually, the predicted differences between the two cuing methods were within the range of differences expected due to inter- or intrapilot variability. It should be noted, however, that the assumptions concerning the information available from a *g*-seat, as well as those concerning the quality of that information, were not as firmly based on data as were the corresponding assumptions for platform motion.

The analyses of the Harrier control tasks, while very informative, are of limited value in demonstrating the ability of the pilot–simulator model to predict performance data in a complex task because there were no empirical data available against which to compare model predictions. The two applications to follow allow for such a comparison.

In a combined empirical and analytic effort of a fairly basic nature, Levison and Junker (1977, 1978) applied the OCM to the analysis of the effects of motion cues on a roll-axis-tracking task. They compared performance when only visual (instrument) cues were available with performance, when the visual cues were augmented, with confirming motion cues. The visual display was an indication of error between target and vehicle roll angle; i.e., it was a compensatory display and did not provide separate indications of the target input and vehicle state. The effects of motion were examined for two task environments: one a disturbance regulation task and the second a target-following task. These two tasks are representative of a disturbance motion condition and a maneuver motion condition, respectively, as described by Caro (1979). Because the frequency content of the motion signals was well within the bandpass of the human motion sensory mechanisms, it was not necessary in this study to model those mechanisms with precision in order to get accurate results. Thus, motion sensing was modeled using an informational analysis approach whereby the ''display'' vector in the OCM was augmented to include the angular position, rate, acceleration, and acceleration rate of the motion base.

Figures 7 and 8 from Levison and Junker (1977) compare model results with experimental measures across the various experimental conditions. In particular, Figure 7 presents rms performance scores for tracking error, plant state variables, and control variables. Also shown is a cost score, which is a weighted sum of error and control variables. In this Figure the static (no-motion) condition is indicated by an S and the motion condition by an M; TARG and DIST refer, respectively, to the target input and disturbance input cases. In Figure 8, subject describing functions and remnant spectra are compared. Excellent agreement between model predictions and experimental data is evident across all the conditions. It must be noted that the model results were obtained with an essentially fixed set of parameters for the pilot model. The only changes made in the OCM between conditions were those made either to change the task description, that is, from target input to disturbance input, or to change the information available to the subject depending on whether it was a motion or a no-motion condition. Both the model and the data show that the addition of motion cues has virtually no effect on performance or pilot strategy for the target input (maneuver motion) condition, but it has a significant beneficial effect for the disturbance motion condition, thus confirming Caro's observations.

Figure 8 shows that when the pilot's frequency response measures are exam-

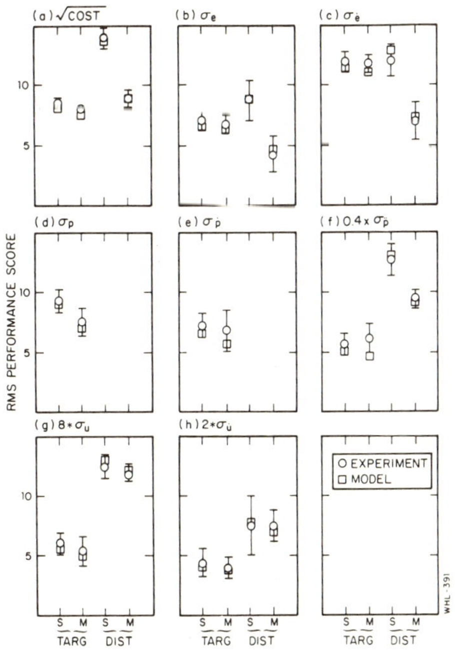

Figure 7. Comparison of OCM predictions and experimental performance scores for roll-tracking task.

ined (describing function gain and phase and remnant level), significant differences due to motion were found for *both* conditions. These effects are summarized in Table 3. If the pilot's gain and phase characteristics are interpreted as a stimulus–response characteristic, and his remnant interpreted as a measure of "noisiness," then the results show that the presence or absence of motion cues significantly affects basic pilot behavior, under both "maneuver" and "disturbance" conditions. In an engineering simulator these differences could lead to erroneous subjective evaluations by a pilot, whereas in a trainer they might result in the subjects learning an inappropriate strategy.

The results can be understood in model-based terms, specifically in relation to the nature of the state estimation process in the two cases. In the target input case, because there are no unknown inputs to the vehicle, the pilot's knowledge of the vehicle dynamics (as encoded in the Kalman estimator) and of his/her own control input allows the vehicle state to be estimated very accurately. This estimate can be used in conjunction with the tracking error provided by the visual display, for estimation of target states. Thus, the skilled subject can use the information provided by the visual display and his/her knowledge of the system to estimate whatever task-related information is necessary and might otherwise be provided by motion cues; so, in a sense, those cues become redundant. In the disturbance case, however, because there is an unknown input to the vehicle, it is

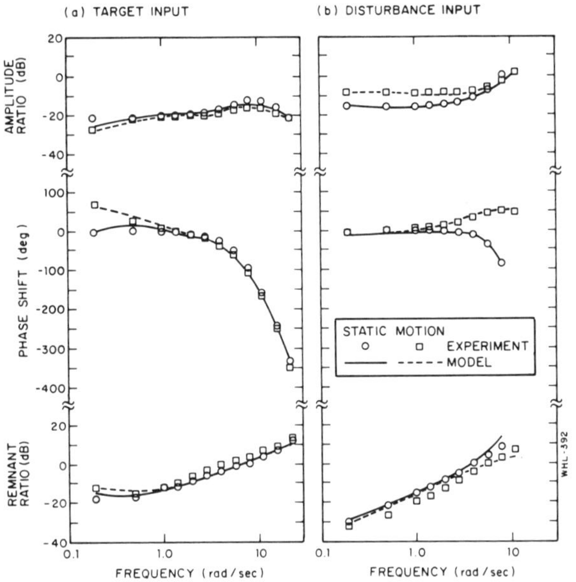

Figure 8. Comparison of OCM predictions and experimental describing functions for roll-tracking task.

not possible to predict the vehicle state on the basis of the control input and vehicle dynamics alone; the additional information concerning the response of the vehicle that is provided by the motion cues is then very important for vehicle state estimation and, in turn, in conjunction with the visual display, for target state estimation.

As a final example of the use of the OCM to investigate the utility of motion cues and a case in which model predictions for a complex task could be compared with experimental results, we consider an analysis of a complex hover

Table 3. Effects of Motion Cues on Frequency Response

	Effects of Motion Cues	
Measurement	*Manuever Condition*	*Disturbance Condition*
Low-frequency phase	Substantial increase in phase lead	No change
High-frequency phase	Small increase in phase lag	Convert phase lag to phase lead, a substantial change
Low-frequency amplitude-ratio	No change	Substantial increase
Gain-crossover frequency	No change	Increase by over factor of 2
Remnant ratio	Overall increase	Overall decrease

control task. Briefly, the simulated flight task corresponded to maintaining a Huey Cobra helicopter in a high hover position relative to a ship moving at 15 knots. The desired hover position was 50 feet from the ship in line with a diagonal deck marking and at a mean height of 20 feet above the deck. The helicopter was subjected to simulated turbulence. Motions of the deck of the ship could also be simulated, and the presence or absence of such ship motion was an experimental variable. The data collected during the experiment were rms values of (rail–rail, bow–stern, and height) tracking errors, helicopter pitch and roll, and positions of the helicopter controls (collective, pitch and roll cyclic, and rudder pedal). The experiment was conducted at NASA Langley Research Center and is described in detail in Ricard et al. (1981).

The pilot–simulator model was specialized to this task by an appropriate definition of the equations of motion for the Huey Cobra, the definition of simulator and perceptual models, and the specification of appropriate human performance parameters. All of the above are described in some detail in Baron (1981b), but three aspects of the formulation deserve special mention. First, when the coupled vehicle dynamics, the rotor dynamics, the SAS, and the gust and ship motion disturbances are all included, there are 36 state variables needed to model the "system" and environment (34 states for the case of no ship motion). Moreover, this does not even include any state variables to account for simulator dynamics or human sensory dynamics. This was a significantly larger problem than any previously treated with the OCM (and, indeed, probably larger than any treated by other manual control models). Second, it was assumed that any dynamics associated with motion cuing or sensing could be neglected so the cue generation and perception problems were treated on an information level. Third, it was assumed that differential delays between motion and visual cues could also be neglected. This last assumption is generally less tenable than the previous one but, as will be seen, does not appear to impact the results significantly.

Twelve experimental conditions were investigated for the main experimental variables of motion cuing (F = fixed base; M = motion base; G = *g*-seat), visual delay (ON, OFF), and ship movement (ON, OFF). Each condition was repeated 5 times by each of 12 pilots, so that there were 60 runs per condition.

The reader is referred to Ricard et al. (1981) for a detailed analysis of the data using univariate analysis of variance. The discussion here will be very brief. Significant effects were found on at least some measures for all the experimental variables, including the replicate and pilot variables. Indeed, the pilot factor was the major source of variance among the experimental variables. Motion cuing and visual delay had significant effects on the error measures, but ship movement did not. Visual delay tended to increase the values of all system variables (error, states, and controls). Platform motion had a significant effect on vehicle roll attitude but not on pitch. Cyclic pitch increased with platform motion, while cyclic roll decreased. Most pilots used more control when the ship motion was active. In addition to the main effects, there were a number of significant interactions, several of them involving the pilot factor. For example, the effects of visual delay was more pronounced when there was no ship motion.

The OCM pilot–simulator model specified for this problem was used to predict the rms values for all the variables of interest in each of the 12 conditions. The model results were obtained with a *fixed set of pilot parameters*—the only changes made in the model from condition to condition were those corresponding to changes in the simulation (e.g., presence or absence of platform motion, *g*-seat, or ship motion). When used in this fashion, the OCM is intended to be representative of the average performance expected of a highly trained, highly motivated pilot who has been allowed to reach asymptotic performance on each condition before data are taken.

Comparison of model results with the data revealed that all measures predicted by the model were within one standard error of the corresponding experimental means, where the standard error was a measure of the variability in the various measures that could be attributed to pilot variability alone. This agreement implies that the model was as good a predictor of a pilot's performance across the conditions as another pilot selected randomly from the group. Given the complexity of the problem and the assumptions made, this is a very encouraging result from the standpoint of human performance modeling.

With a few exceptions, the model predictions were also within one standard deviation of the mean (where the variation was across both subjects and replications) for virtually all measures and all conditions. Even the exceptions were well within two standard deviations of the mean, and most were only marginally more deviant than one standard deviation. A sample of these results illustrating the degree of agreement between model and data is provided in Figures 9 and 10, where rail-to-rail position error and pitch-angle error are plotted.

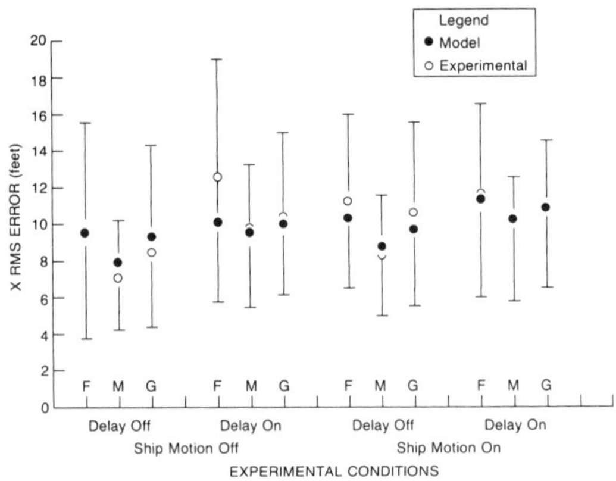

Figure 9. Comparison of OCM predictions and data for Huey Cobra hover performance (rail-to-rail error).

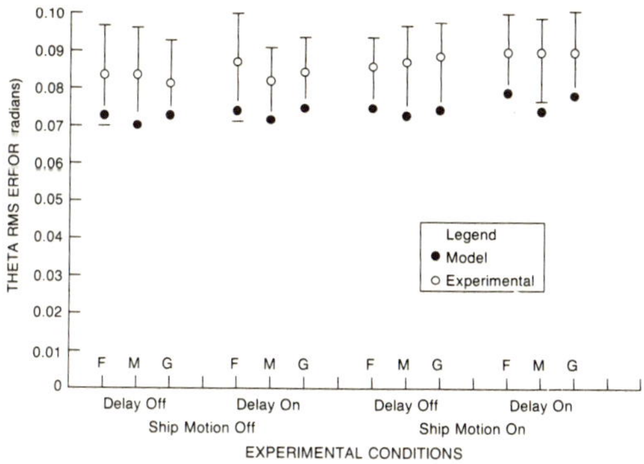

Figure 10. Comparison of OCM predictions and data for Huey Cobra hover performance (pitch error).

Examination of the model results as a function of the main experimental variables is also of interest here. Figure 11 shows the effect of motion cues on rail-to-rail error (as predicted by the model), averaged over ship motion conditions, for the two delay conditions. It can be seen that the model suggests that *g*-seat cues will provide some improvement in performance and platform motion cues will result in still less error. Further, the effect of motion cues do not appear to be significantly different for the two delay values; i.e., there does not appear to be a delay–motion interaction. These effects of motion cues exactly parallel the trends contained in the data (see fig. 11 of Ricard et al., 1981).

The effect of motion cues on the attitude variables (ϕ and θ) are shown in Figure 12. For this figure, model results for each motion condition were averaged across the delay and ship motion variables. The *g*-cues are predicted to have only a very slight effect on the attitude variables. The provision of platform motion, however, should result in demonstrably lower rms attitude values, with

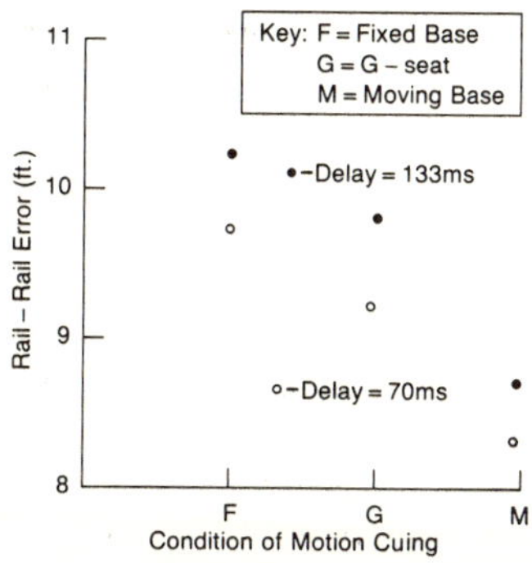

Figure 11. Effects of delay and motion condition on OCM predicted Huey Cobra hover performance (rail-to-rail error).

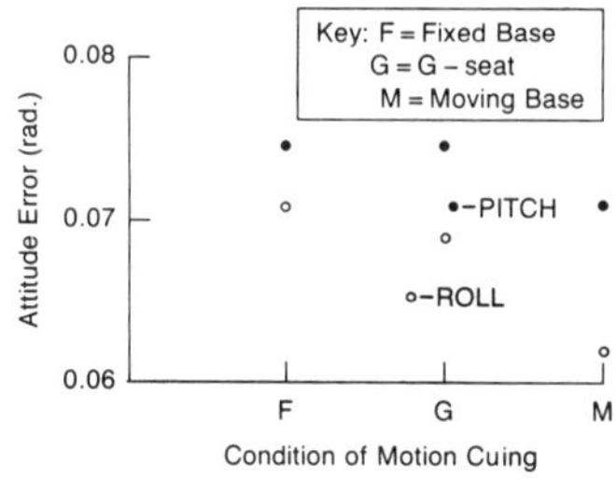

Figure 12. Effect of delay and motion condition on OCM predicted Huey Cobra hover performance (attitude errors).

the effect for roll being greater than that for pitch. Again, these effects closely parallel those in the data where there was a significant effect of motion on roll, but not on the pitch rms score.

The model did less well in predicting the effects of motion cues on control inputs. In addition, while the OCM predicted the trend of performance degradation with increased visual delay, the model tended to underestimate the effects in the data; i.e., the OCM was doing a better job of compensating for the delay than were the pilots.

In summary, the model predictions agreed with the data to a substantial degree, with most of the differences being within the range of experimental variability. There appeared to be a general trend for the model predictions to be less sensitive to configuration changes than the actual measures of mean performance. This may be expected because the model assumes asymptotic performance for each condition and this may not be achieved in the experimental situation. The model did particularly well in predicting the effects of motion cues on position errors and attitude angles. This suggests that the simple informational treatment of motion cues employed in the study (and the values preselected for parameters of the model) were adequate for prediction of performance in this task. The main effects of visual delay were also predicted by the model, as was the lack of motion–delay interaction. However, the model predicted less degradation with delay than was observed in the ''mean'' rms performance measures.

2. *Visual Cues*

Thus far, the discussion has centered on applications of the model to the analysis of motion cues. Although motion cuing requirements have received a great deal of attention in the literature, and there still remain fundamental questions concerning them, the simulator community has started to devote considerably more attention to questions associated with simulating the external visual environment. Many flight control tasks depend upon such an exterior view of the visual world. These include critical flight tasks such as landing under VFR conditions, nap-of-the-earth flight, weapon delivery, air-to-air combat, air refueling, and formation flying.

As noted in Section III, there have been several studies aimed at developing OCM-based models for the pilot's use of extracockpit visual cues in control tasks. Here, we limit our discussion to two studies: the first (Wewerinke, 1979), discussed very briefly, was a basic study to test the ability to incorporate external visual cues in the OCM; the second (Baron, Lancraft, and Zacharias, 1980) is closely tied to simulator fidelity and is indicative of the possibilities provided by the model-based analysis.

Wewerinke (1978, 1979) performed basic studies of a pilot's use of external visual cues in approach to landing, concentrating on specific geometric visual cues for the task. Because the view of the runway is by far the most useful visual cue in a conventional VFR approach, Wewerinke was able to use highly schematized line drawings to study the effects of the various cues provided by that environment. Within this context, he considered the pilot's use of field orientation cues for attitude information, linear perspective for positional information, and motion parallax cues for self-velocity information. Through the use of elementary perspective geometry, Wewerinke related changes in orientation of the basic line segments comprising the extracockpit visual scene to rotational and translational vehicle movements, i.e., to vehicle state variables. In this way, he was able to specify an appropriate set of "display variable relations" for use in the OCM.

An issue in Wewerinke's investigations was the question of whether or not the pilot's use of such external visual cues could be represented adequately within the OCM framework and, if so, whether a reasonable perceptual model could be derived that would allow for predictive use of the model in future applications. For this to be the case, one would have to be able to obtain good matches to the empirical data based on reasonable assumptions concerning the OCM and its parameters, including those relating to the perceptual model (in particular, the observation noise-to-signal ratios, the visual thresholds, and the attention-sharing parameters).

Wewerinke found that the OCM could provide a fairly close match to the rms score data based on a reasonable set of assumptions for the model. In particular, he fit that data with approximately "nominal" pilot-related parameters consistent with earlier OCM analyses of symbolic display tasks and a set of a priori values for visual thresholds that were inferred from a related but independent psychophysical experiment. Upon closer examination of the data, however, he found that the frequency domain measures of pilot randomness (i.e., the observation noise measures) were not in good agreement with those predicted by the model. He was able to match these measures without significantly degrading the score predictions by allowing the perceptual thresholds to vary. These thresholds differed somewhat from those based on the psychophysical experiments and may be more relevant for modeling the perception process in tracking tasks.

There were a number of limitations to the studies conducted by Wewerinke (see Zacharias and Levison 1980, for a discussion). However, it remains the most basic and extensive combined empirical and analytic study of its kind to date. It showed not only how the OCM could be used to predict performance using external-visual-cue-type information, but also how that model could be employed to infer what information in a scene was most necessary to the task and how the pilots were making use of that information. The results also demonstrate the importance of a wide range of performance measures for developing an accurate picture of how the pilot is processing the information available.

In a more direct and wider-ranging study of simulator fidelity, Baron, Lancraft, and Zacharias (1980) used the OCM model of Figure 1 to explore the effects of changes in the parameters of a computer-generated image (CGI) visual system and a platform motion system on performance in a helicopter hover task. The assumed task was to hover over a fixed point at a fixed altitude, in the presence of disturbances generated by air turbulence. The specific helicopter chosen for this study was the CH-47 tandem rotor transport helicopter. Visual cuing was provided by a CGI system that simulated a out-the-window view, and motion cuing by a six-degree-of-freedom motion simulator (VMS).

An OCM pilot–simulator model corresponding to that in Figure 1 was developed for this task and is described in detail in the reference. Here, we discuss briefly some of the key assumptions and then examine how the model was used to analyze simulator fidelity.

It was assumed that the equations of motion for simulating vehicle dynamics were implemented on a digital computer, operating at a nominal update rate of 30 Hz, and, for simplicity, that the only effect of digitization was the introduction of a sample and hold delay associated with the base cycle time of the mainframe computer.

The CGI characteristics of interest in the analysis are listed in Table 4, along with assumed nominal values for those characteristics. Off-nominal CGI characteristics were also considered. These included variations in the refresh rate, the effective sample rate, the field of view, and the display resolution.

Each axis of the motion system was modeled as a second-order dynamic system with appropriate position–rate–acceleration servo limits. Table 5 defines the nominal parameter values associated with each motion axis. The nominal

Table 4. CGI Characteristics

Picture Refresh Rate	30 frame/sec
Display Compute Time	66 msec
Effective Sample Rate/Delay	15 Hz/99 msec
Scene Content	6000 edges/frame
Field of View	3 screens across (144° horiz, 36° vert)
Display Resolution	1024 lines/frame × 1024 pixels/line
No CGI Lag Compensation	

Table 5. Motion System Characteristics

Linearized Transfer Function: $\dfrac{\lambda}{\lambda_c} = \dfrac{\omega_n^2}{s^2 + 2\zeta_n\omega_n s + \omega_n^2}$

Axis	$\omega_n{}^a$	ζ_n	$\ddot{\lambda}^{max}$	λ^{max}	λ^{min}	λ^{max}
Roll (ϕ)	9.4	0.7	50°/sec^2	15°/sec	−22°	22°
Pitch (θ)	9.4	0.7	50°/sec^2	15°/sec	−24°	26°
Yaw (ψ)	9.4	0.7	50°/sec^2	15°/sec	−29°	29°
Surge (x)	9.4	0.7	16 ft/sec^2	2 ft/sec	−2.5 ft	2.5 ft
Sway (y)	18.8	0.7	24 ft/sec^2	10 ft/sec	−20 ft	20 ft
Heave (z)	18.8	0.7	32 ft/sec^2	20 ft/sec	−30 ft	30 ft

[a]Rad/sec.

motion system did not include washout filters, as all predicted motions except the surge (x) motion were well within their respective simulator limits.

A simplified visual perception model was used for the analysis. It can be shown that the six vehicle position coordinates (three rotational and three translational) can be related geometrically to azimuth and elevation line of sight (LOS) angles to objects in the visual field. The essence of the simplified approach was to assume that the pilot could infer the position variables directly from the objects in the visual field but that the accuracy in doing so would be limited. These limitations reflect constraints on resolving or detecting changes in the LOS angles (arising from either human or CGI-imposed resolution limits) and those arising because of the necessity for shifting visual attention so as to pick an "optimal" object for inferring a particular vehicle state. Consequently, in the OCM formulation it was assumed that (1) the vehicle position variables were "displayed" directly; (2) the pilot's visual thresholds depended, in part, on the geometry relating object–LOS angles to position coordinates and, in part, on inherent perceptual limitations; and (3) the pilot had to share attention among the "displays" of vehicle position.

Perception of motion cues was based on the model obtained from the review of Zacharias and noted in Section III. This model was simplified to reduce computational requirements by identifying passbands that accounted for the majority of the correlated power coming out of the vestibular system and eliminating any lead-lag elements having break frequencies not in the passbands.

The hovering task was modeled as a disturbance regulation task. As is standard procedure for application of the OCM, it was assumed that the objective of the task could be characterized as minimization of a cost function of the following form:

$$J = E[y_i/y_{i_{max}})^2 + r_j\dot{u}_j{}^2],$$

where $y_{i_{max}}$ is a performance tolerance on the corresponding variable. The variables of interest y_i consist of vehicle positions (x, y, z) and velocities $(\dot{x}, \dot{y}, \dot{z})$ and attitude (ψ, θ, ϕ) and attitude rate $(\dot{\psi}, \dot{\theta}, \dot{\phi})$ variables. The weightings on control rate activity, r_j, were chosen by means of an error–control trade-off analysis to yield appropriate values for the neuro motor time constants (T_N) associated with the OCM (see Figure 1). The reader is referred to the reference for specific details.

The OCM, as described above, was used to analyze the effects of CGI and motion system limitations on closed-loop hover performance, with the goal of determining the effects of simulator characteristics on performance and workload. Baron, Lancraft, and Zacharias addressed the issue of simulator fidelity directly by defining a "perfect" or ideal simulator in which there were no simulation time delays, no motion system dynamics, and an infinite-resolution imagery system. Essentially, this simulator configuration corresponded to the flight condition itself. Model predictions for this idealized configuration provided a benchmark against which to measure simulator deficiencies. To explore the fidelity question over a range of conditions, six basic simulator configurations were defined so that the effects of the visual and motion systems could be evaluated both separately and together. These configurations are listed in Table 6.

Given that results can be obtained from the model for an assumed flight condition, a useful relative performance metric for the task can be defined. For each plane of control (i.e., longitudinal and lateral), relative performance was defined by the following relationship:

$$\text{Relative performance (in \%)} = 100 \times (J - J_{flt})/J_{flt}$$

where J is the value of the cost functional defined above for the configuration under investigation and J_{flt} corresponds to the value of J predicted by the model

Table 6. Simulator Configurations

Configuration	*Description*
Perfect (Flight)	No simulator delays, nominal field of view, human operator thresholds, no VMS dynamics
Perfect CGI–Realistic VMS	Includes main frame computer delays and VMS platform dynamics
Realistic CGI–Perfect VMS	Includes main frame and display computer delays, CGI imposed visual thresholds, no platform dynamics
Realistic CGI–Realistic VMS	Includes all simulator nominal characteristics
Perfect CGI–Fixed Base	Includes main frame computer delays, no motion cues
Realistic CGI–Fixed Base	Includes CGI limitations, no motion cues

for flight (the perfect simulator). Relative performance is a model prediction of the percentage deviation from "flight" performance introduced by the particular simulator characteristics of interest and is therefore a relevant metric for simulator fidelity. The use of J for performance comparison (rather than individual error and control scores) is advantageous because this quantity is a scalar metric of overall performance and therefore provides a concise summary of the simulator effects. In addition, this type of combined measure of performance has been suggested for measuring training performance (Waag et al., 1975). The value of J may also be correlated with vehicle flying qualities (Hess, 1977), which suggests that increases in J owing to simulator deficiencies may lead to erroneously degraded "flying qualities" evaluations by pilots for the simulator as compared to the actual vehicle.

Figure 13(a,b) shows relative performance predictions of the model for five simulator configurations relative to that expected from the perfect simulator (which, by definition, has a relative performance of zero). It can be seen that the changes in simulator configuration have different effects for the lateral and longitudinal control axes. For the longitudinal control task, the CGI fidelity is the most significant factor. Indeed, performance is predicted to be better with a perfect CGI and no motion than with perfect motion and a realistic CGI. However, motion is still important, particularly when there is a realistic CGI system, as can be seen by comparing performance for the cases involving a realistic CGI. In the lateral control task, the results are similar, but motion is even more important. In this case, having a perfect CGI does not compensate for lack of motion. Moreover, adding motion ameliorates the consequences of visual deficiencies.

The above model predictions are based on the assumption of a fixed observation noise ratio. As we have noted previously, this model parameter can be related either to the attention devoted to the task or to the skill level of the pilot. Figure 14 shows how relative performance on the lateral control task changes as a function of observation noise ratio for several simulator configurations. The solid dots on the curves indicate the performance obtained with the nominal level of observation noise ratio assumed. It can be seen that the relative ordering of simulator configurations is maintained at all values of observation noise ratio. At low values for this parameter, i.e., at high levels of attention or at very high skill levels, the performance with the realistic CGI-perfect VMS configuration approaches that for the realistic VMS-perfect CGI configuration. Apparently, if the noise/signal ratio is lowered significantly on the motion cues, it can offset some of the visual deficiencies associated with the realistic CGI. The above results suggest that the visual processing limitations introduced by a nominal CGI configuration could result in significant deterioration of closed-loop hover performance.

It is of major interest to the designer to determine the effects of variations in individual design-related CGI parameters. This was accomplished in this study by first relating OCM parameters to CGI design factors as indicated in Figure 15 and then computing performance for the appropriate set of model parameters. In

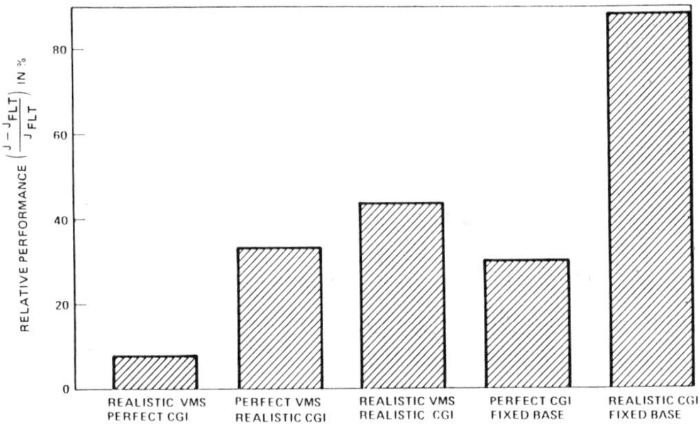

Figure 13a. Relative performance vs. simulator configuration (CH-47 longitudinal axis).

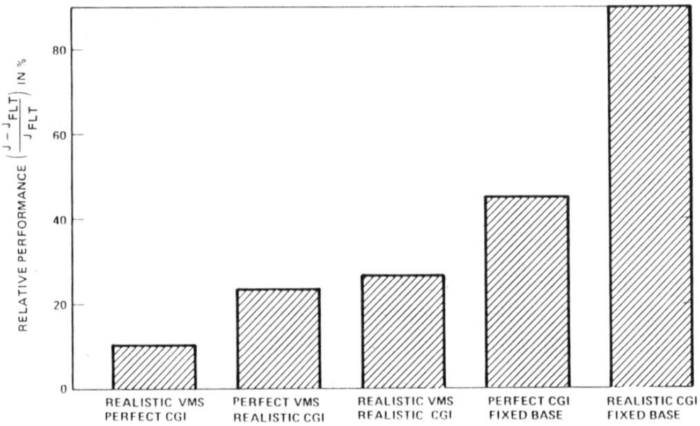

Figure 13b. Relative performance vs. simulator configuration (CH-47 lateral axis).

this way, Baron, Lancraft, and Zacharias were able to examine the effects of simulator time delay, CGI field of view and display resolution, and motion system bandwidth on performance. Without getting into the details, the results indicate that, with respect to the CGI system, the most important parameter in terms of its effect on performance was display delay. This was followed in order of importance by display resolution and field of view.

The general trends of the results for this study were fairly consistent across conditions. However, the magnitude of the effects and the relative importance of various parameters clearly depended on the specific task, as exemplified here by the differences between longitudinal control and lateral control. Moreover, there were significant interactions between visual and motion cuing conditions, i.e.,

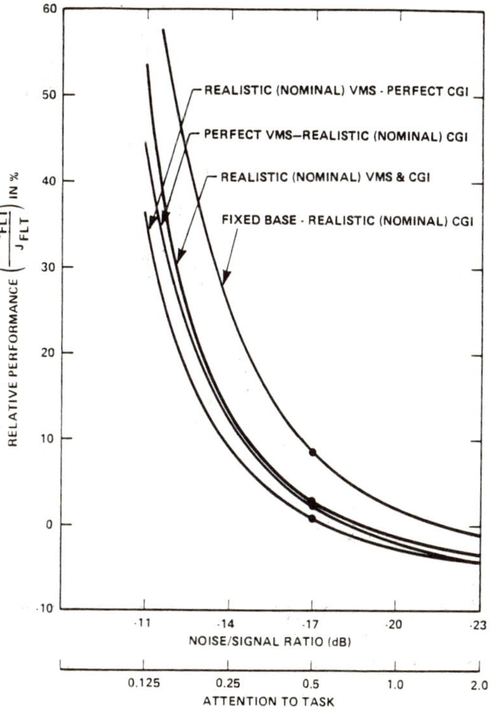

Figure 14. Effect of attention on relative performance for different simulator configurations (CH-47 lateral axis).

the effect on performance of degrading or improving one set of cues was a function of the quality of the other set of cues. These factors illustrate the complexity of the simulator fidelity question, but they also show the potential for models of the pilot simulator system to address those questions.

B. Analysis of Simulator Modeling Requirements

Another major aspect of simulator fidelity concerns the modeling of the aircraft characteristics. This aspect of providing faithful flight characteristics is sometimes neglected in comparison to cue fidelity or is included as part of providing adequate cues. Of course, the problem of obtaining suitably accurate data on the aircraft is a difficult one that is receiving more attention with the procurement of new aircraft. However, even if perfect data were available to model the aircraft and its systems, the simulation computers would invariably introduce distortions and/or discrepancies. With modern simulators, incorporating digital computers for computation of aircraft response, a major problem is computational delay, though other problems also exist.

Baron, Muralidharan, and Kleinman (1980) examined the effects on closed-

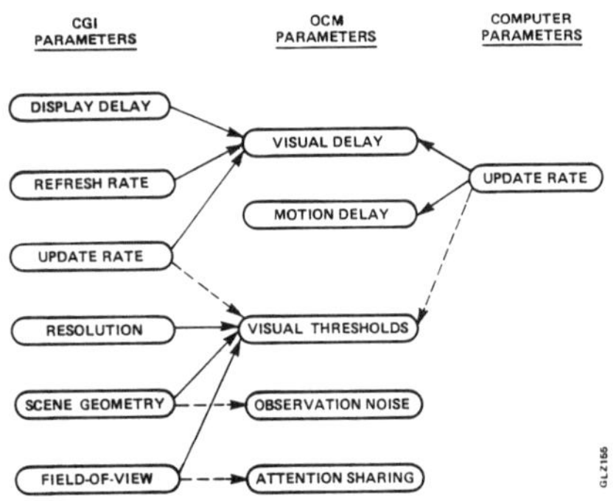

Figure 15. Relation of simulator characteristics and OCM parameters.

loop pilot simulator performance of a number of factors that introduce distortions into the simulation. They developed models that could be incorporated in the OCM for various components of a digital flight simulator. Then, using the OCM, they explored the effects on closed-loop performance of changes in simulator parameters. In particular, a longitudinal target tracking task was explored and results were obtained for two digital integration schemes (Euler and Adams-Bashforth, AB2), for two mainframe sample rates (32 Hz and 16 Hz), for two bandwidths (25 rad/sec and ∞) for the servo projectors used to provide a visual horizon and for two control loader configurations (analog and digital).

An experiment was performed at NASA Langley Research Center (Ashworth, McKissick and Parrish, 1979), the data from which were used to test the predictions of the closed-loop model. The experimental data were analyzed using analysis-of-variance methods. This involved averaging across conditions and subjects, while holding one (main) experimental variable fixed. Thus, for example, to compare Euler integration with Adams-Bashforth, *all* runs for each integration scheme were averaged, including those at different sample rates, visual delays, etc. The results of this analysis are shown in the fourth column of Table 7, which provides a comparison of experimental results with those obtained from the OCM pilot–simulator model. Presented there are the average tracking error for each of the "main" conditions and the standard deviation for all runs. The "best" condition corresponds to Adams-Bashforth integration at 32 Hz with no visual servo delay and an analog control loader implementation; the "worst" condition corresponds to Euler integration at 16 Hz with the 25 rad/sec visual servo bandwidth and the discrete control loader. It should be noted that the "best" simulator condition had the least simulation delay and the "worst" condition the greatest delay, though delay was not the only difference.

The model predictions for the various conditions were averaged to mimic the statistical analysis of the experimental data. Thus, e.g., to obtain the result for Euler integration, the separate model scores obtained for the eight conditions

Table 7. Comparison of Model and Experimental Results

	RMS Vertical Tracking Error (ft)(Mean ± 1 Std. Dev.)		
Configuration	*Model Fixed Parameters*	*Model Single Parameter "Fit"*	*Experiment*
Euler	10.76 ± 0.74	12.69 ± 2.50	12.67 ± 3.19
AB2	9.67 ± 0.68	9.67 ± 0.68	9.69 ± 1.59
16 Hz	10.72 ± 0.81	12.66 ± 2.60	12.57 ± 3.11
32 Hz	9.69 ± 0.65	9.69 ± 0.65	9.79 ± 1.89
25 rad/sec Visual Servo	10.59 ± 0.79	11.55 ± 2.33	12.26 ± 2.90
Infinite Bandwidth Visual Servo	9.83 ± 0.83	10.81 ± 2.41	10.10 ± 2.52
Digital CL	10.49 ± 0.91	11.45 ± 2.43	11.35 ± 2.83
Analog CL	9.93 ± 0.6	10.91 ± 2.36	11.01 ± 3.02
Worst	11.69 ± 0.52	15.50 ± 0.46	16.27 ± 2.08
Best	8.80 ± 0.28	8.80 ± 0.28	8.39 ± 0.97

involving Euler integration were averaged. Similarly, to obtain results for 32-Hz sample frequency, the eight runs in which a 32-Hz sample rate was assumed were averaged. And so on.

The results of this analysis for fixed model parameters are shown in the second column of Table 7. The model trends are quite similar to those exhibited in the data. For this case, however, the model results are somewhat optimistic with respect to the "poorer" simulator conditions. In addition, the model predicts less variation between conditions than exists in the data. We expect this to be the case, to some extent, inasmuch as the model assumes ideal experimental controls such as asymptotic training levels on all conditions and consistent motivation across conditions.

In an attempt to obtain closer agreement to the data and to the variation across conditions, a single model parameter, the observation noise/signal ratio, was allowed to vary for one (the worst) simulator condition. In particular, a noise/signal ratio, was allowed to vary for one (the worst) simulator condition. In particular, a noise/signal ratio of −11 dB was assumed for the most difficult case (i.e., Euler integration and 16-Hz sample frequency). With this one assumption, the results shown in the third column of Table 7 and in Figure 16 were obtained. These results are in excellent agreement with the data with respect to both specific values and trends. Moreover, the model results now show a variation much more like that of the data. It is not clear just why the higher noise/signal ratio should be associated with the poorest simulator conditions. Two possible explanations are (1) the information environment is degraded by

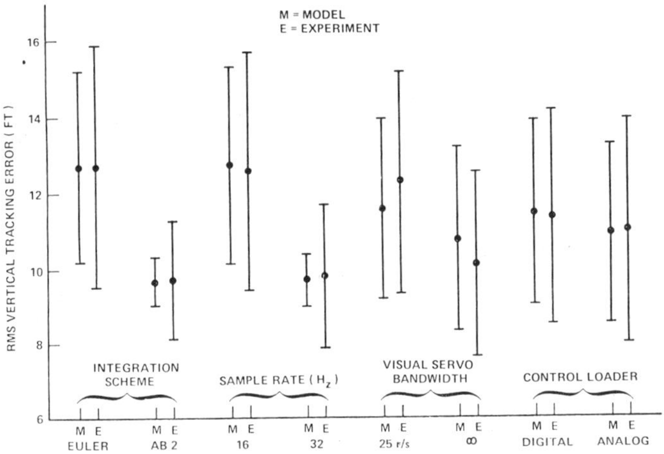

Figure 16. Comparison of OCM predictions and experimental tracking error for different simulator models.

the simulator delays (in a nonlinear way) to the point where the operators find it much more difficult to obtain the required information and this is reflected in the present model in a higher observation noise; or (2) because of the delay, the task becomes so difficult and prone to pilot-induced oscillations (PIO)s that the pilots do not control as tightly or as continuously and this is reflected by a lower level of attention and therefore a higher noise. This issue clearly deserves further analysis.

C. Development of Model-Centered Fidelity and Effectiveness Metrics

The earlier discussions of simulator fidelity and of modeling approaches to assessing simulator effectiveness suggest the potential value of a simulator fidelity metric that reflects the following considerations:

1. The metric should be centered on human performance capabilities and limitations, with a greater emphasis on the human's sensory–perceptual, cognitive, and psychomotor performance and a lesser emphasis on the specific physical characteristics of the simulator.
2. The metric should support more than just qualitative generalizations concerning major categories of simulator requirements; instead it should be capable of providing objective and quantitative assessments of simulator fidelity, assessments which can be directly translated into detailed simulator design specifications.
3. The specific objectives of the simulation, such as the mission–task training requirements should contribute to determining the metric value assigned to a particular simulator design.

The previous results and discussions indicate that it may be possible to meet these requirements with quantitative OCM-centered metrics that exhibit the desired human performance relevance and the ability to incorporate simulator objectives directly. We elaborate these points below.

1. *Human Performance Relevance*

The OCM, with its perceptual (noise–delay estimator), control law, and "neuromotor" dynamics "blocks," neatly parallels behavioral categories of sensory–perceptual processing, cognitive behavior, and psychomotor performance. Assuming that the OCM is a reasonable functional analog of the human (acting under the constraints imposed by a task requiring a high degree of proficiency), then the model provides us with a means of dissecting human behavior and examining the interactions between major behavioral components of skilled performance.

As we have seen, Levison and Junker were able to model in detail the human's use of motion cues in disparate tasks with an integrated, uniform approach that involved simply augmenting the set of perceptual variables available to the modeled operator to include the position, rate, acceleration, and acceleration rate of the roll axis simulator. This simple "information augmentation" was reasonable in light of the above-threshold and within-bandpass stimuli used in their experiments. When the stimuli are relatively low-level, or when there is the chance of distortion due to the dynamics of the sensory end organ, threshold and dynamic modeling of the sensor is called for. The work of Borah et al. and Baron, Lancraft, and Zacharias show that this is also relatively straightforward within the framework provided by the OCM.

The central estimator structure of the OCM provides a convenient and appealing means with which to integrate findings in the physiological literature concerning sensory end organ characteristics with current knowledge of closed-loop manual control. In this way, the OCM acts to unify the sensory–perceptual and manual control research areas and provides a solid foundation for understanding human performance differences due to the different cuing environments typically found in an aircraft and its associated simulator.

2. *Quantitative Fidelity Metrics*

In contrast to many other approaches that might be used to define a fidelity metric, the use of the OCM provides the potential for the definition of objective and quantitative fidelity metrics, a feature that will become increasingly more important as simulator design questions grow in subtlety and complexity. The above discussions have indicated the nature of the performance metrics the OCM is capable of providing (performance scores, describing functions, and remnant levels), as well as the accuracy with which model predictions of these quantities match the experimental data. We have also seen, in the study of Baron, Lancraft, and Zacharias, how the OCM might be used to define a specific fidelity metric, relative performance, and, in turn, how this metric might be applied to evaluating simulator fidelity. The general approach is summarized below.

We presume that we have a specific aircraft and task to simulate. We also presume that we have a simulator for this aircraft and task, and the objective is to assess its fidelity for training purposes.

The first step in developing a quantitative fidelity metric would be to analyze and model skilled pilot performance for the "flight" case. This would involve a computer simulation with the OCM, using a set of "standard" human performance parameters to predict suitable pilot performance metrics associated with control of the actual vehicle. Note that none of the constraints imposed by the simulator would be operative here. Note, too, that a broad range of performance measures could be considered. For example, metrics based on rms station-keeping error, rms control stick input, pilot describing functions, and pilot remnant are possible. The measures predicted by the OCM for these conditions would define "criterion" pilot performance in the actual vehicle task.

The next step would be to conduct a similar analysis of the simulator-based task, again using the OCM to predict (simulator) pilot performance measures. One would expect these predictions to differ from those obtained in the first analysis because of simulator–vehicle differences. For example, the simulator motion system (if there were one) might be characterized by long delays and/or low bandwidths; or the extracockpit scene might have a low update rate typical of early CGI designs.

The third step would consist of combining the relevant performance measures, from both analyses, into a single fidelity metric which is a function of both vehicle and simulator pilot performance. This fidelity metric, call it F, could be a combination of component metrics based on scores, etc. (Baron, 1985). This third step, then, would result in the computation of a value for F, the OCM-based fidelity metric. This value would be the guide to evaluating the simulator's fidelity in terms of the desired objectives: a small value would indicate the simulator's appropriateness for the task, and a large value would indicate serious simulator deficiencies.

At present, it is not possible to correlate the fidelity value F directly with some absolute scale of simulator "goodness" or "badness"; instead, the metric would provide a relative measure of simulator effectiveness, useful for comparing one simulator against another or for assessing the impact of potential simulator design changes. This use of the metric in a "differential" fashion also would directly support the definition of simulator design requirements.

The discussion to this point has illustrated the definition and use of a model-based metric by concentrating on performance equivalence. The OCM can also be used to develop a metric based on perceptual fidelity, which could then be used in a similar fashion as the performance metric. This is so because the OCM provides predictions, or analogs, of internal signals related to perception. For example, one might define a metric based on the error covariance of the estimates of the state variables. This covariance can be directly related to the efficiency in extracting information about the state from the available measurements (Baron and Levison, 1977). For the same estimation strategy, lower errors imply that the information available is better. The relative difference in these errors between simulator and flight is a measure of the simulator's ability to reproduce the perceptual environment for the pilot. Unfortunately, such a metric is primarily an analytic tool inasmuch as the pilot's estimate (or the error in it) cannot be measured directly.

It is perhaps less clear that metrics based on (attentional) workload fidelity are also possible. This is accomplished by examining the sensitivity of performance to changes in observation noise/signal ratio for various simulator configurations (Baron, Lancraft, and Zacharias, 1980).

Finally, a hybrid metric, based on performance equivalence, perceptual fidelity, and workload fidelity, could be defined. A weighted combination of the metrics could be used to ensure that performance, workload, and perception were reflected in the simulator fidelity metric, with the relative importance of each dictated by the simulator objectives.

3. *Relevance to Simulator Objectives*

The discussion has thus far centered on development of quantitative metrics and their human performance relevance. It is important to examine how simulator objectives are considered in this context. The approach can be applied to objectives for both engineering and training applications of simulators. The relation of the metrics to training objectives is the more difficult problem and is discussed here.

The types of training objectives specifications best suited to use with the OCM are those which specify the desired performance levels to be achieved at the end of simulator training. These performance levels could be specified relative to skilled in-flight performance or could be specified as absolute levels. The previous discussions have shown how the OCM could be used to generate predictions of in-flight performance of skilled pilots that could serve as a training objective. An example of a specification based on absolute levels of performance might be as follows: "A student will be able to maintain a vertical hover error of less than 3 feet." Inasmuch as the actual situation is more probabilistic than that implied by a hard-and-fast performance criterion, one might translate the specification to read along the following lines: "A student will maintain a vertical hover so as not to exceed an error of 3 feet more than 1 percent of the time." A statistical performance measure such as this would then be used directly in constructing the performance metric for the OCM. One could then determine the fidelity required so that this performance level is achievable in the simulator.

The methodology is *not* restricted to considering one training objective at a time (as might be inferred from the above example) and, in fact, is well suited for handling the multidimensional problem associated with the simultaneous specification of a number of performance objectives imposed on the pilot trainee. Thus, desired performance might also involve a specification of "control smoothness," level of pilot "randomness," and/or changes in control strategy.

The OCM-based fidelity metrics can be related to broader measures of simulator effectiveness for training in a fairly straightforward manner. Consider an evaluation based on performance fidelity. Suppose a specific simulator allows a trainee to achieve an asymptotic performance level P^{sim} (say, for vertical hover). If a skilled pilot can achieve a level $P^{a/c}$ in the actual aircraft, then one might define the following performance fidelity metric:

$$F = \left(\frac{P_{\infty}^{\text{sim}} - P_{\infty}^{a/c}}{P_{\infty}^{a/c}} \right),$$

so that a zero F value would reflect perfect fidelity, and large values would reflect deficiencies in the simulator. If one simply defined the simulator's effectiveness E as inversely proportional to the product of its cost C and its fidelity level F,

$$E = 1/C \cdot F)$$

then one would have a quantitative means for comparing different simulators in terms of their effectiveness. Thus, a low-cost but poor-fidelity simulator might rate higher in effectiveness than a high-cost, high-fidelity one, depending on the relative differences in cost and fidelity.

This provides an initial estimate of effectiveness but it is incomplete because it does not account for the hourly cost (in dollars and trainee hours) of training to a given criterion level on the simulator. If CPH is the simulator cost per hour and T_{sim} is the time needed to train to the criterion level, then the above effectiveness measure could be amended to

$$E = 1/(C \cdot F) + 1/(\text{CPH} \cdot T_{\text{sim}}).$$

Thus, a simulator with a high hourly cost would be penalized in its effectiveness rating, unless the high cost bought efficient training and, with it, a short time to train to the criterion level.

This type of effectiveness measure requires some estimate for the training time T_{sim}. A rough approximation could be obtained by assuming a certain level of information-processing capability for the naive trainee and then using the model to predict his/her performance at the beginning of training. With another model-based estimate for asymptotic performance at the end of training, an exponential learning model could then be fit to these end points. The intersection of this model curve with the criterion level would provide an estimate of T_{sim} for use in the above relation.

To assess more accurately the impact of simulator configuration on training efficiency, one would have to predict or estimate T_{sim} on the basis of a model of learning behavior. Such a learning model does not presently exist in the pilot-vehicle closed-loop control context, but there has been some preliminary work (Levison, 1983). Finally, it should be pointed out that the above discussion presumes that training objectives can be defined in terms of performance objectives to be attained in the simulator as compared to flight. It does not directly address the transfer question; i.e., the model does not predict initial performance in the aircraft given predicted asymptotic performance in the simulator. To evaluate something like the cumulative transfer effectiveness ratio (Roscoe, 1971), which is of interest in the training community, one would have to predict this initial transfer performance level as well the rates of learning in both simulator and aircraft. We are not aware of any serious efforts to develop a predictive model for transfer of training.

D. Design of Compensation for Simulator Delay

Thus far, we have examined how the OCM could be used to evaluate simulator fidelity. In this section, we consider the use of the model in improving fidelity through the design of schemes that compensate for simulator deficiencies. In particular, we illustrate here how the model may be employed to develop filters that compensate for undesirable delays in simulators. The reader is referred to appendix B of Baron, Lancraft, and Zacharias (1980) for a discussion by Levison on the use of the OCM for designing control laws (washout filters) to drive simulator motion bases.

Unwanted delays in digitally based flight simulators are inherent and arise from the basic sample-data nature of the simulation, the algorithms employed, and the lags generated in simulating the appropriate cue environment. With very-high-speed, powerful computers used to compute aircraft response, the latter source of delays tends to dominate, particularly with high-resolution computer-generated imagery being employed. It can be shown theoretically (Baron, Kleinman, et al., 1970; Baron, Muralidharan, and Kleinman, 1980) and has been demonstrated in manual control experiments (Queijo and Reilly, 1975; Ricard and Harris, 1980; Levison, 1979; Crane, 1980) that the presence of transport delays in wide-band, precision control tasks will degrade performance. Moreover, it is also recognized that, in these situations, delays can increase pilot workload and alter pilot control strategy (Ricard and Harris, 1980; Crane, 1980). For these reasons, it is important to minimize simulation hardware and software delays in the design process and subsequently to compensate for any remaining significant delay, if possible.

Of course, the significance of a given simulation delay will depend on the particulars of the simulated flight task, i.e., on the vehicle dynamics, disturbances, and control objectives. Moreover, the appropriate form and parameters for a delay compensation filter will also depend on these factors. These complications, along with the multivariable nature of the flight control problems of prime interest today, make it difficult, if not altogether impractical, to design compensation filters on a "rule-of-thumb" basis.

Baron, Lancraft, and Caglayan (1982) conducted an analytic study to explore the application of the OCM to the delay compensation problem. The basic view underlying their approach was that the objective of delay compensation for piloted simulation is to restore the situation, as seen and experienced by the pilot, as closely as possible to that which would be obtained in the absence of unwanted delay. Because perfect compensation is not realizable (it requires a pure time advance), one can only hope to accomplish this objective approximately. There are several ways in which the OCM can be applied to the delay compensation problem. The most obvious approach is to predict the effects of delay on the performance of the system and the pilot (including pilot workload). Then, candidate compensation filters can be evaluated using model predictions, and the parameters of those filters may be "tuned" to improve their performance. Less obvious is the use of the model for direct design of compensation filters. Two approaches to applying the OCM directly to compensator synthesis were investigated by Baron, Lancraft, and Caglayan. Both approaches appeared to lead to very effective compensators, particularly for delays on the order of those associated with current CGI systems. However, one of the approaches was easier to apply and generalized more readily to multivariable systems.

In this chapter, results for the compensation problem for a single-loop roll-tracking task are presented and discussed. (The reader is referred to the cited report for application of the techniques to a multiinput, multioutput helicopter simulation). The roll-tracking task is illustrated in Figure 17.

The vehicle dynamics were a low-order approximation to those used by Crane (1980) in an analytical–empirical study of delay compensation that utilized a describing function approach to pilot modeling. The input was chosen to yield a display change that approximated the vehicle's response to a wide-band disturbance.

The roll-tracking task was analyzed with the OCM using standard parameters for the pilot model. A computational delay of 0.018 sec and a display delay varying from 0 to .096 sec was assumed, to correspond to the experimental situation described in Crane (1980). Mean-squared errors and pilot and open-loop describing functions were computed as a function of delay with the OCM. The results for the error scores are shown in Figure 18, which compares computed and experimental effects of delay on performance scores. The experimental scores are the average of five pilots, each flying eight trials at each condition (Crane, 1980). Both the experimental and computed scores are normalized by their respective zero-delay values so that a comparison of the relative effect of delay may be made. As can be seen, model results predict the effect of delay on tracking error quite well.

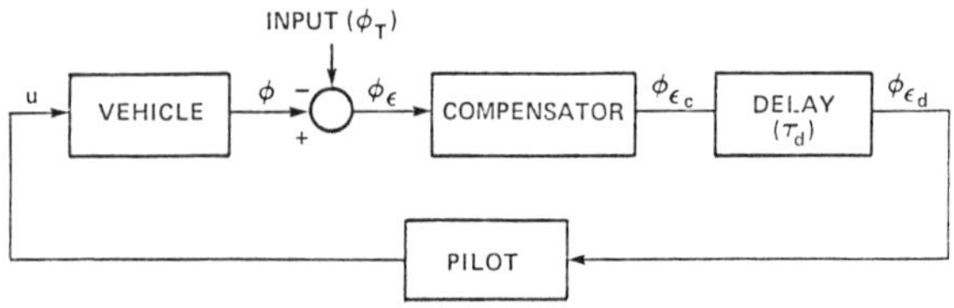

Figure 17. Block diagram for roll-tracking task.

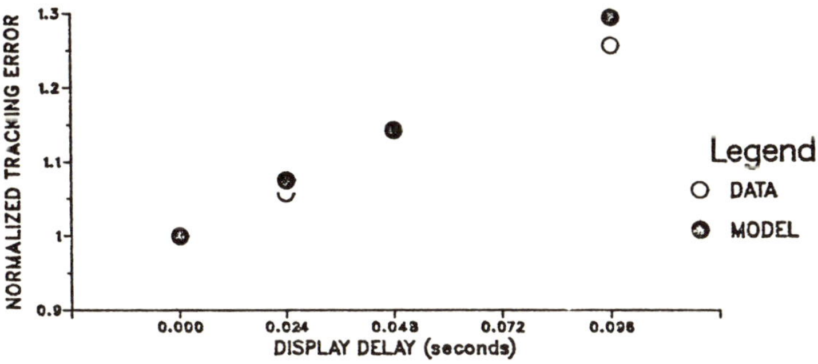

Figure 18. Effect of delay on roll-tracking error.

Three different delay compensators were designed and were tested analytically: a classical compensator, an OCM frequency domain compensator, and an OCM time domain compensator. The design approaches leading to these compensators are described in detail in Baron, Lancraft, and Caglayan (1982). Briefly, the classical compensator was based on Crane's approach and involved choosing the parameters of a lead-lag filter in an attempt to restore performance and phase margin to the no-delay values. The OCM frequency domain compensator utilized a rational transfer function approximation to the OCM optimal predictor for the given delay; the idea was to relieve the pilot of the necessity for generating the additional lead (prediction) required to compensate for simulator delay. The above two approaches have a critical shortcoming when one attempts to extend them to multiinput, multioutput (MIMO) problems. The third compensator, the OCM time domain compensator, was a filter designed by state-space methods that could deal directly with the MIMO problem. The basic idea behind the design of this compensator was to use the OCM to develop an analytic expression that could be used as a fixed lead predictor for the *closed-loop* pilot–simulator dynamics.

Figure 19 shows the effect each compensation scheme has on the pilot describing functions predicted by the model. The pilot describing function is defined as the transfer function from y to u through the OCM pilot model, and as such, defines the adopted pilot "strategy" for the various configurations. The phase portion of Figure 19 shows that with no compensation, but with full display delay (bullet), the pilot must introduce (a maximum of) approximately 20° more phase lead at $\omega = 7$ rad/sec than for the no-delay case (box). All three compensators reduce the requirements for added pilot phase lead. However, both of the OCM-based compensators appear to provide a better match to the desired no-delay phase than the classical compensator. The corresponding gain curves show that, at frequencies up to about 4 rad/sec, all compensators yield a predicted pilot gain that lies between that for the no-delay and full-delay cases. Moreover, all the gain differences in this region are relatively small (<2.5 dB). Larger discrepancies are apparent in all compensated cases at the higher frequencies, where the

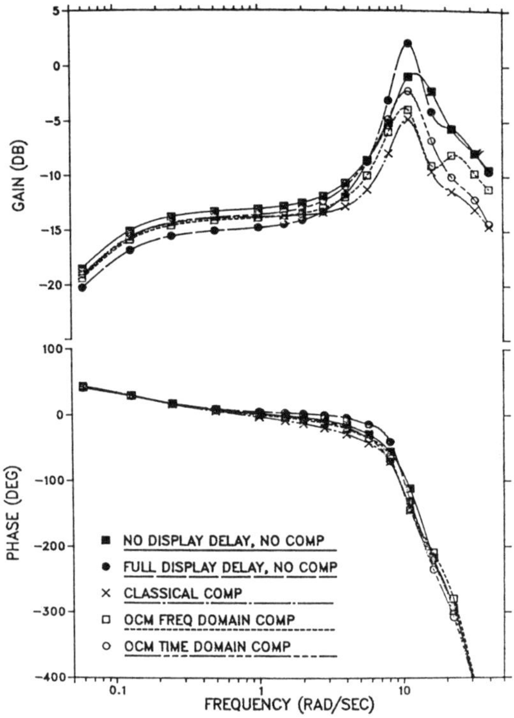

Figure 19. Effect of compensation on predicted pilot-describing function (roll-tracking task).

effects on performance of gain mismatch are less important. Over the entire frequency range of interest, the OCM time domain compensator (ring) provides the closet match to the no-delay pilot gain, even doing the best job of mimicking the resonance characteristic in the region around $\omega = 10$ rad/sec.

The ability of a compensator to generate a match to various performance levels found in the no-delay situation is another important indication of the compensator's effectiveness. Table 8 displays the rms scores predicted by the OCM for each condition. Included in the table are scores for roll error ϕ_e, roll rate error $\dot{\phi}_e$, pilot control u, and control rate $\dot{u}$. In addition, the last column contains computed values for a cost functional J that is a weighted sum of roll error variance and control rate variance. Table 8 shows a 25 percent decrease in performance when the display delay is present, but no compensation is used (as noted in Figure 18). Inclusion of any of the compensators, as designed, results in a restoration of performance back to no-display delay levels (with a corresponding decrease in pilot workload). Inasmuch as lowering J is tantamount to matching roll error and control rate variances, we see that, whereas all compensators can match roll error equally well, the OCM time domain compensator does significantly better in matching control scores.

Table 8. Computed RMS Error Scores vs. Condition—Roll-Tracking Task

	Variable				
Condition	ϕ_{ϵ_c}	$\dot{\phi}_{\epsilon_c}$	u	$\dot{u}$	J
No Delay, No Compensation	3.82	15.0	1.29	9.84	16.8
Full Delay, No Compensation	4.34	15.2	1.29	9.88	21.0
Classical Compensator	3.82	17.9	1.02	7.36	15.8
OCM Frequency Domain Compensator	3.81	14.5	1.06	7.45	15.7
OCM Time Domain Compensator	3.82	18.7	1.21	8.82	16.3

Based on the single-axis results presented above, the OCM time domain compensator was selected as the compensation scheme best able to replicate the desired no-display delay conditions. This, however, is not to say that the other methods were without merit. On the contrary, in many instances, they performed just as well as the selected scheme. However, the ability of the OCM time domain compensator to be extended relatively easily to the MIMO case was seen as a very useful feature. The results for applying the compensator to such a case were reported in Baron, Lancraft, and Caglayan; they were quite encouraging and suggest that empirical testing of the method is desirable.

V. CONCLUDING REMARKS

The application of a pilot–simulator model to the analysis of issues related to simulator effectiveness, particularly simulator fidelity, has been examined. Attention was restricted to high-value flight control tasks. The OCM-based model was described along with some relevant models for sensory perception. Then, several applications were considered to illustrate the utility and scope of the model for addressing important questions in a quantitative fashion.

In the author's view, the results reported herein suggest that the state of modeling human control behavior is such that the pilot–simulator model described above can be used to analyze many aspects of simulator effectiveness today in a manner that is both objective and valuable. However, there are some limitations that constrain application and suggest needs for further research and development.

First, it should be stated that the pilot–simulator model is just one piece of the total picture and its value would be enhanced significantly by integrating it into an overall simulation effectiveness modeling system. This is especially true for application to the development of training simulators where what is needed most are analytic tools that allow one to perform cost-effectiveness trade-offs among training objectives, alternative training technologies and trainer use, *prior to* actual trainer development. Such a system should comprise several component models (e.g., human performance models of cognition and control, machine models of reliability and costs) to permit the objective prediction of costs and effectiveness of alternate trainer designs.

The pilot simulator model and the data bases that support it also have some deficiencies that currently limit the potential applications. Further development is needed in the sensory area with respect to extracockpit visual cuing, proprioceptive cuing, and tactile cuing. Both models and data are needed in these areas; the problems are being addressed to a limited degree in current research. With respect to compensation for simulator deficiencies, the major need at this time is for experimental testing and evaluation of compensators designed by model-centered approaches. In the training context there are several areas in which the modeling technology is deficient. The most important of these are inadequate models for skill acquisition, very little attention to or modeling of individual differences, and lack of a model for predicting transfer of training. Thus, in general, the existing models can provide reasonable predictions of asymptotic performance after training on a device (or the vehicle), but presently there are no validated models to predict how long it will take to achieve that level of performance or how early performance and learning in a second "device" (the aircraft) is affected by training on a different one (the simulator).

On the other side of the ledger, there are a number of simulator-related uses for human performance models that have not been discussed. Models of the type described in this paper have been, and are being, developed for tasks other than flight control, e.g., failure detection and monitoring (Baron, 1985). Thus, it should be possible to utilize analytic pilot models for a much wider range of flight tasks than addressed herein. An intriguing use of a pilot model would involve embedding it in the simulator to assist in instructional support functions. Such functions generally involve instruction, monitoring and evaluation of performance, control and individualizing of training, and playing the role of a potential adversary or controller of the aircraft. In each instance, one can can imagine how an appropriate model for the human operator might be used. For instruction purposes, a normative model might well be used to demonstrate how the task should be performed. Such a model would also have value in terms of monitoring and evaluating training performance. In particular, one might compare the performance levels attained by the trainee against criterion performance levels predicted by the normative model, or one might even compare strategies of the trainee with those of a normative model. Such monitoring and evaluation could then provide the basis for controlling training events, i.e., for an adaptive training program. Finally, a human operator model might be used to control an adversary such as in air-to-air combat training, or it might be used as a controller of the trainee such as, for example, in a model for the LSO officer to be used in training carrier landing.

ACKNOWLEDGMENTS

This paper grew out of an effort to examine the role of models in assessing simulator effectiveness that was supported by the Naval Training Equipment Center under Contract Number N6133983C1030. Dr. Gil Ricard was the contract monitor for the effort. The underlying work owes much to present and former colleagues at BBN, particularly Bill Levison, Greg Zacharias, Roy Lancraft, Ramal Muralidharan, and Alper Caglayan. In addition, other past sponsors of the work include USAF–AMRL (particularly Andy Junker, Grant McMillan, and Ed Martin), NASA Langley Research Center (particularly Roland Bowles, Russ Parrish, and Burnell McKissick), and NASA Ames Research Center (particularly Frank Crane).

NOTES

1. The design of a "trainer" would involve additional considerations such as instructional modes and performance measurement systems, but these are not the subject of this chapter.
2. Of course, engineering judgment or more formal analyses will frequently be used to simplify the model significantly. For example, if signal frequency content is well within the bandwidth of a system, the dynamics of that system may be ignored; this is often the case for vestibular dynamics. Such simplifications are very useful in reducing computational costs in applying the OCM and have been used extensively in past studies (see below).
3. The model for attention sharing is used to modify the basic observation noise covariances—the less attention devoted to a variable, the higher the corresponding observation noise.
4. By "motion cues" we mean sensation of motion as provided by the pilot's vestibular sensory organs or by proprioceptive or tactile cues and not motion as might be inferred from visual perceptions.

REFERENCES

Ashworth, B. R., McKissick, B. T. and Parrish, R. V. 1979. The effects of simulation fidelity on air-to-air tracking. *Proceedings of the Fifteenth Annual Conference on Manual Control, Wright State University, Dayton, Ohio.*

Baron, S. 1980. *A Multi-Cue OCM for Analysis of Simulator Configurations.* Cambridge, MA: Bolt Beranek and Newman Inc., Report No. 4373.

Baron, S. 1981a. *A Pilot/Vehicle Model Analysis of the Effects of Motion Cues on Harrier Control Tasks.* Navtraequipcen 80-D-0014-1, Naval Training Equipment Center, Orlando, FL.

Baron, S. 1981b. *An Optimal Control Model Analysis of Data from a Simulated Hover Task.* NAVTRAEQUIPCEN 80-C-0055-1, Naval Training Equipment Center, Orlando, FL.

Baron, S. 1984. Adaptive behavior in manual control and the optimal control model. In O. G. Selfridge, E. L. Rissland, and M. A. Arbib (Eds.), *Adaptive Control of Ill-Defined Systems.* New York: Plenum, pp. 51–73.

Baron, S. *Human Performance Models for Assessment of Simulator Effectiveness.* Cambridge, MA: Bolt Beranek and Newman, Inc., Report No. 5869.

Baron, S. and Berliner, J. E. 1975 *MANMOD 1975: Human Internal Models and Scene-Perception Models.* Technical Report No. RD-CR-76-3, U.S. Army Missile Command, Redstone Arsenal, AL.

Baron, S. and Berliner, J. E. 1977. *The Effects of Deviate Internal Representations in the Optimal Model of the Human Operator.* TD-CR-77-3, U.S. Army Missile Research and Development Command.

Baron, S. and Kleinman, D. L. 1968. *The Human as an Optimal Controller and Information Processor*. NASA CR-1151.

Baron, S., Lancraft, R., and Caglayan, A. 1982. *An Optimal Control Model Approach to the Design of Compensators for Simulator Delay*. NASA CR-3604, National Aeronautics and Space Administration, Ames Research Center, CA.

Baron, S., Lancraft, R., and Zacharias, G. 1980. *Pilot/Vehicle Model Analysis of Visual and Motion Cue Requirements in Flight Simulation*. NASA Contractor Report 3312, NASA Ames Research Center, CA.

Baron, S. and Levison, W. H. 1977. Display analysis with the optimal control model of the human operator. *Human Factors 19*(5): 437–457.

Baron, S. and Levison, W. H. 1980. The optimal control model: status and future direction. *Proceedings of the International Conference on Cybernetics and Society, Cambridge, MA* pp. 90–101.

Baron, S., Muralidharan, R., and Kleinman, D. L. 1980. *Closed Loop Models for Analyzing Engineering Requirements for Simulators*. NASA 1-14449.

Baron, S., Pew, R. W., and Zacharias, G. L. 1980. *Performance Measurement for Simulator Evaluation*. Cambridge, MA: Bolt Beranek and Newman, Inc., Report No. 4315.

Borah, J., Young, L. R., and Curry, R. E. 1978. *Sensory Mechanism Modeling*. Wright Patterson Air Force Base, OH: Air Force Human Resources Laboratory, AFMRL TR-78-83.

Caro, P. W. 1979. The relationship between flight simulator motion and training requirements. *Human Factors 21*(4): 493–501.

Carter, V. E. and Semple, C. A. 1976. *Specific Behavioral Objectives for VF-121 Training in Basic Air Combat Maneuvers,* Vol. 3. NOR 76-52.

Crane, D. F. 1980. Time delays in flight simulator visual displays. *Proceedings, Summer Simulation Conference, Seattle, WA*.

Grunwald, A. and Merhav, S. J. 1974. *An Analytical Model for Visual Field Control: A Parametric Study*. Haifa: Israel Institute of Technology, Report TAE-217.

Gum, D. R. 1973. *Modeling of the Human Force and Motion Sensing Mechanisms*. Wright Patterson Air Force Base, OH: Advanced Systems Division, Air Force Human Resources Laboratory, AFHRL-TR-72-54.

Heffley, R. K., Clement, W. F., Ringland, R. F., Jewell, W. F., Jex, H. R., McRuer, D. T. and Carter, V. E. 1981. *Determination of Motion and Visual System Requirements for Flight Training Simulators*. Technical Report 546, U.S. Army Research Institute for the Behavioral and Social Sciences.

Hess, R. 1977. Prediction of pilot opinion ratings using an optimal pilot model. *Human Factors 19,* vol. 5, pp. 459–475.

Kleinman, D. L., Baron, S., and Levison, W. H. 1970. An optimal-control model of human response, Part 1: Theory and validation. *Automatica 6:* 357–369.

Levison, W. H. 1983. *Development of a Model for Human Operator Learning in Continuous Estimation and Control Tasks*. Cambridge, MA: Bolt Beranek and Newman, Inc., Report No. 5331.

Levison, W. H., Elkind, J. I., and Ward, J. L. 1971. *Studies of Multivariable Manual Control Systems: A Model for Task Interference*.'' NASA CR-1746, National Aeronautical and Space Administration, Ames Research Center, CA.

Levison, W. H. and Junker, A. M. 1977. *A Model for the Pilot's Use of Motion Cues in Roll-Axis Tracking Tasks*. AMRL-TR-77-40, Aerospace Medical Research Laboratory, Wright Patterson Air Force Base, OH.

Levison, W. H. and Junker, A. M. 1978. A model for the pilot's use of motion cues in steady-state roll-axis tracking tasks. *Proceedings of the AIAA Flight Simulation Technologies Conference, Arlington, TX*.

Levison, W. H., McMillan, G. R., and Martin, E. A. 1984. Models for the effects of g-seat cuing on roll-axis tracking performance. *Proceedings of the 20th Annual Conference on Manual Control,* NASA Ames Research Center, CA.

Levison, W. H. and Zacharias, G. L. 1980. An optimal control model for the joint use of visual and motion cues in continuous tracking tasks: theory and experiments. *Journal of Cybernetics and Information Sciences* (Special Issue on Man–Machine Systems) *3*(1 4).

Levison, W. H., Zacharias, G. L., and Sinacori, J. B. 1982. *Design of an Experiment to study the Pilot's use of Visual and Motion Cues in a Height-Regulation Task.* Cambridge, MA: Bolt Beranek and Newman Inc., Report No. 5028.

Matheny, W. G., et al. 1971. *An Investigation of Visual, Aural, Motion and Control Movement Cues.* Naval Training Devices Center, Report NAVTRADEVCEN 69-C-0304-1.

McRuer, D. T. and Krendel, E. S. 1974. *Mathematical Models of Human Pilot Behavior.* AGARD Report AGARD-AG-188.

McRuer, D. T., Weir, D. H., Jex, H. R., Magdeleno, R. E., and Allen, R. W. 1973. Measurement of driver/vehicle multi-loop response properties with a single disturbance input. *Proceedings of the Ninth Annual Conference on Manual Control,* MIT, Cambridge, MA.

Pew, R. W., Baron, S., Feehrer, C. E., and Miller, D. C. 1977. *Critical Review and Analysis of Performance Models Applicable to Man–Machine Systems Evaluation.* Cambridge, MA: Bolt Beranek and Newman Inc., Report No. 3446.

Queijo, M. J. and Riley, D. R. 1975. *Fixed-Base Simulator Study of the Effect of Time Delays in Visual Cues on Pilot Tracking Performance.* NASA TN D-8001.

Ricard, G. L. and Harris, W. T. 1980. Lead/lag dynamics to compensate for display delays. *Journal of Aircraft 17*(3).

Ricard, G. L., Parrish, R. V., Ashworth, B. R., and Well, M. D. 1981. *The Effects of Various Fidelity Factors on Simulated Helicopter Hover.* NAVTRAEQUIPCEN 1H-321, Naval Training Equipment Center, Orlando, FL.

Roscoe, S. N. 1971. Incremental transfer effectiveness. *Human Factors 13*(6): 561–568.

Rouse, W. B. 1980. *Systems Engineering Models of Human Machine Interaction.* Amsterdam and New York: North-Holland Publ.

Semple, C. A. 1981. *Simulator Training Requirements and Effectiveness Study (STRES): Executive Summary.* AFHRL-TR-80-63, Wright Patterson Air Force Base, OH.

Sheridan, T. B. and Ferrell, R. 1974. *Man–Machine Systems.* Cambridge, MA: MIT Press.

Waag, W. L., Eddowes, E. E., Fuller, J. H., Jr., and Fuller, R. R. 1975. *ASOPT Automated Objective Performance Measurement System.* AFHRL-TR-75-3, Brooks Air Force Base.

Wewerinke, P. H. 1978. A theoretical and experimental analysis of the outside world perception process. *Proceedings of the Fourteenth Annual Conference on Manual Control.* NASA Conf. Pub. 2060.

Wewerinke, P. H. 1979. Visual scene perception: frequency domain and model parameter estimation procedure. *Proceedings of the Fifteenth Annual Conference on Manual Control.* AFFDL-TR-79-3134, Air Flight Dynamics Laboratory, Wright Patterson Air Force Base, OH.

Zacharias, G. L. 1978. *Motion Cue Models for Pilot–Vehicle Analysis.* USAF AMRL-TR-78-2.

Zacharias, G. L., Caglayan, A. K., and Sinacori, J. B. 1983. A model for visual flow-field cueing and self-motion estimation. *Proceedings of the 1983* American Control Conference,

Zacharias, G. L., Caglayan, A. K., and Sinacori, J. B. 1983. A visual cueing model for terrain-following applications: *Proceedings of the AIAA Flight Technology Simulation Conference, Niagara Falls, NY.*

Zacharias, G. L. and Levison, W. H. 1980. *Development of a Model for the Use of Extra-Cockpit Visual Cues.* Cambridge, MA: Bolt Beranek and Newman Inc., Report No. 4652.

Zacharias, G. L. and Young, L. R. 1977. Manual control of yaw motion with combined visual and vestibular cues. *Proceedings of the 13th Annual Conference on Manual Control. MIT, Cambridge, MA.*

AUTHOR INDEX

SUBJECT INDEX